THE FOSSMILL STORY

Life in a railway lumbering village on the edge of Algonquin Park

The Fossmill Story

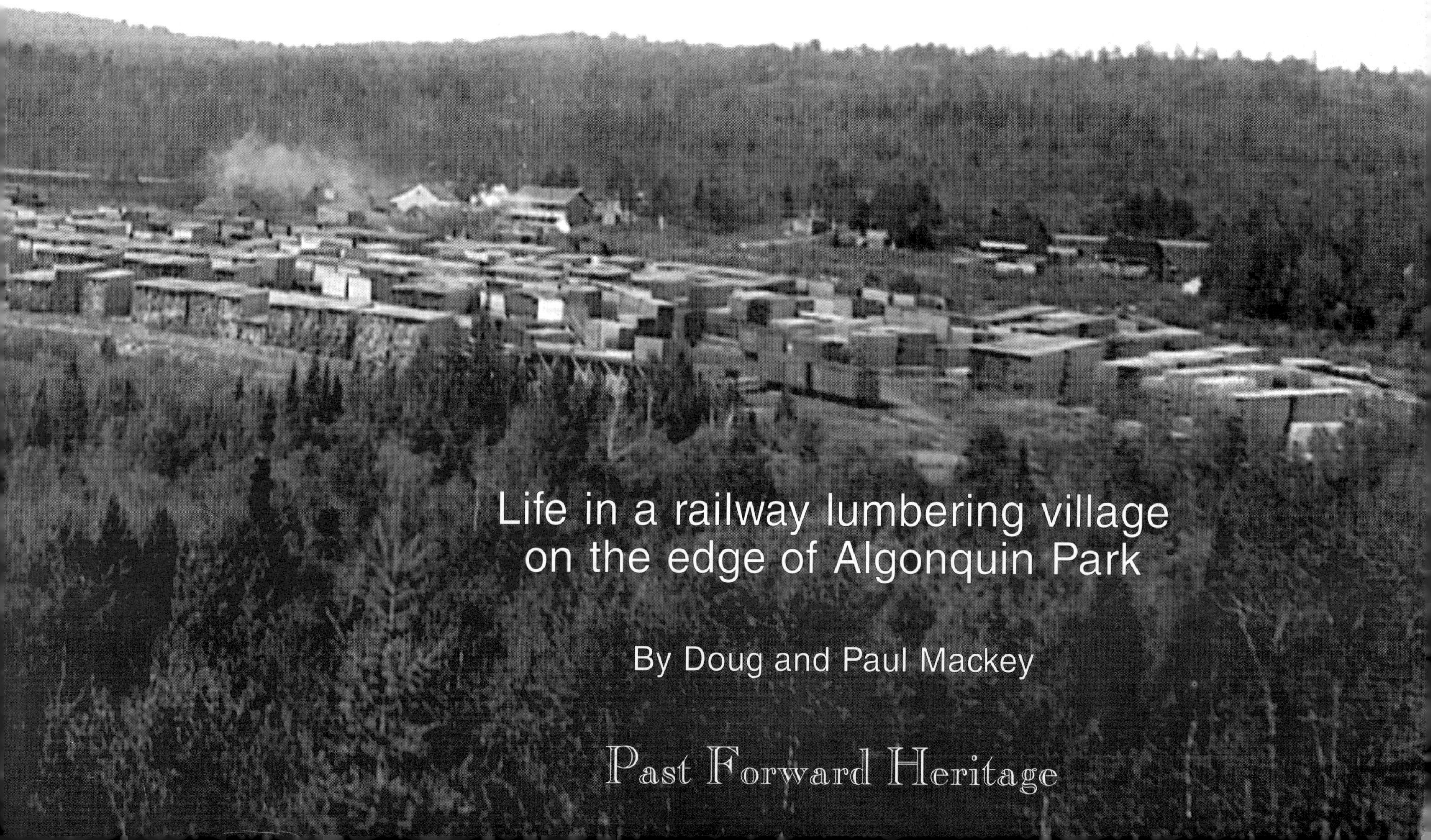

Life in a railway lumbering village
on the edge of Algonquin Park

By Doug and Paul Mackey

Past Forward Heritage

All information and material herein is believed to be accurate. Corrections or new information would be appreciated and may be included in future editions.

Published by Past Forward Heritage
514 Bear Mountain Rd.
RR #1 Powassan, Ontario P0H 1Z0

Canadian Cataloguing in Publication Data

Mackey, Doug, 1931-

The Fossmill story: life in a railway lumbering village on the edge of Algonquin Park

Includes bibliographical references and index.

ISBN 1-896974-10-4

1. Fossmill (Ont.)-- History.

2. Lumbering--Ontario--Fossmill--History--20th century.

I. Mackey, Paul, 1954-

II. Title.

FC3099.F674M33 1999 971.3'147 C99-900835-8

F1059.5.F68M33 1999

Cover, colourization of cover photos, illustrations, and book design by Paul Mackey.

Cover photo: The Denis children in front of the Fassett Lumber Corporation's mill yard Fossmill, Ontario in the late 1920s. For family profile see page 210.

Title page photograph: The Fassett Lumber Corporation mill, Circa 1930 from the Chisholm Woman's Institute.

Past Forward Heritage acknowledges with gratitude the financial support of the Friends of Algonquin Park in the preparation of this publication.

Printed and bound in Canada by University of Toronto Press.

This book is dedicated to the people who lived this story and walked us, in words, through the village, mill, and woods, and shared their privileged memories with us.

Here as I take my solitary rounds
Amidst thy tangling walks and ruined grounds,
And many a year elapsed, return to view
Where once the cottage stood, the hawthorn grew.

—The Deserted Village, *by Oliver Goldsmith (1770)*—

I have been pressing my elbows on the arms of my chair, and dreaming that I was standing on the bridge in front of Dorlcote Mill, as it looked one February afternoon many years ago.

—The Mill on the Floss, *by George Eliot (1860)*—

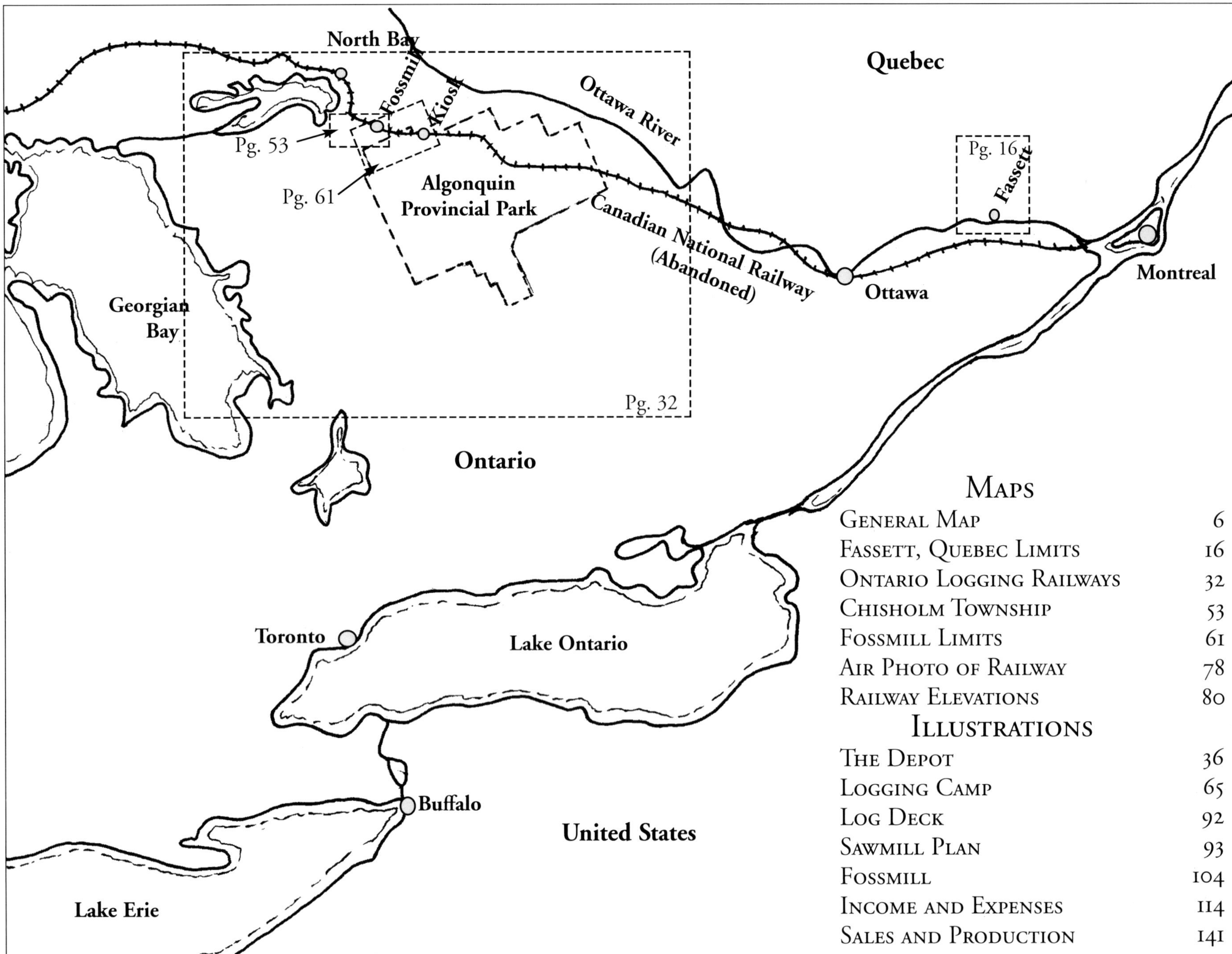

Maps

General Map	6
Fassett, Quebec Limits	16
Ontario Logging Railways	32
Chisholm Township	53
Fossmill Limits	61
Air Photo of Railway	78
Railway Elevations	80

Illustrations

The Depot	36
Logging Camp	65
Log Deck	92
Sawmill Plan	93
Fossmill	104
Income and Expenses	114
Sales and Production	141

Contents

Acknowledgements 8
Introduction 9

The Story

Prologue: Day of Rest 11
1. The Beginning 15
2. A New Home 45
3. The Logging Season 59
4. The Fossmill Railway 77
5. The Lumber Mill 89
6. The Village 101
7. The People 117
8. Fire and Depression 137
9. 1934 151
10. On Relief 163
11. New Hope 175
Epilogue: A New Home, Again! 195

Appendix A: Railway Equipment 203
Appendix B: Relief Payment System 209
List of Interviews 212
Glossary 214
Notes 216
Bibliography 217
Photo Credits 219
Index 220
Places to see and things to do 222

Sidebars

The Honourable J. Sloat Fassett 16
A Ride on the Salmon River & Northern Railway 22
Trout Creek Logging Railway 33
South River Logging Railway 34
Early Logging in Chisholm 35
Canadian National Railway 37
The Fossmill Outlet 40
Jobber's Contract 63
Doyle Scale Rule 69
Cutting Methods 92
Hand Signs 93
Grading Lumber 94
Sawmill Wage Rates 98
Ted Priest 120
Joseph G. Demeza 122
William Mackey and J. R. Booth 172
Sydney J. Staniforth 200
Locomotive Specifications 206
The Denis Family 210

Acknowledgements

As people interested in local and oral history, the authors are fortunate to live a short distance from the former site of the Fassett Lumber Corporation's Fossmill operation, and the village that existed during the first half of the century on the edge of Algonquin Park.

A second piece of good fortune was the fact that the Woman's Institute of Chisholm Township, where the mill was located, had included a brief history of the operation in their two excellent local history books *Our Chisholm Story, Volume 1 and 2.* The third piece of good luck was that we found some local people with intimate knowledge of those years. Stanley Anderson, for example, wrote vivid profiles of Chisholm life in the two Chisholm history books, worked at Fossmill for a period of time, and followed its story with interest (Stanley, age 96, received a 1998 Ontario Heritage Foundation Award for his work on heritage preservation).

Stanley directed us to Stella McGibbon the 98 year old wife of Jack McGibbon, the former Fossmill woods superintendent, and to their daughter Margaret "Peggy" Brushey, who lives nearby in the town of Powassan. A one-hour taped interview with Stella in May 1995 led to the decision to attempt to recreate Fossmill (Stella died 120 days later before we could talk to her again). Peggy went to Fossmill with her parents in 1924, attended school there and was a keen observer of Fossmill life from the management of the operation through all aspects of its evolution and demise. Several lengthy interviews with her set the stage for numerous other interviews throughout the area and the province. Her recollection of a film of the operation led to the acquisition that materialized into the production of a video on Fossmill.

Early interviews with Margaret Frederick, the daughter of the locomotive engineer Allan Gleason, led to contact with her sisters Nora and Fran and brother John, all of whom added to the story. John brought insight into the work and technology of Fossmill. Subsequent interviews with two other original workers, Donat Huneault and Louis Demers, established the basic structure of the story. Numerous other interviews (see page 212) provided details, perspectives and varying points of view on all aspects of Fossmill before, during, and after the Fassett Lumber Corporation's tenure.

Donald and Robert Staniforth, the remaining sons of Sidney Staniforth, the key player in the Fossmill story, co-operated and supported us and provided critical and invaluable information, archival materials, photographs, and film. Film footage on the Fossmill woods and mill operation in the early 1930s was made available to us by Donald Staniforth and became the core of the 1996 video "Logging by Rail in Algonquin Park". The accumulated archival material on Kiosk that came from the Fossmill study will form the basis, in co-operation with former Kiosk residents, of a millennium publication and video on Kiosk, ending a story that started a century before in Fassett, Quebec.

As the interviews progressed, dozens of Fossmill photographs emerged from closets and basements and the story became not only an oral history but along with the video, a visual one as well. Photographs and documents were also acquired from the Algonquin Park Archives, the Ontario Archives, the Federal and Provincial Air Photo Libraries, and The Thomas Fisher Rare Book Library. A special mention is made to Sandra Langlands and the staff of the Sigmund Samuel Library (U of T Library), and the staff of the University of Toronto Archives.

The newspapers, *Canada Lumberman*, and the *North Bay Nugget* were valuable resources. Numerous other sources provided information and confirmations.

Many individuals in the institutions above were helpful, as were others too numerous to mention. A special thanks to Ron Tozer in his old and new roles at Algonquin Park and to Pat Tozer and the Friends of Algonquin Park, who were supportive in many ways. The financial assistance of the Friends of Algonquin Park is especially appreciated and was critical to the completion of the book.

Thanks to the itinerant and knowledgeable Brian Westhouse, who directed us to some important sources, to Ken Cooper, who provided information on the Standard Chemical Company in South River, and to Melody Chapman, who very capably transcribed numerous tapes of text as the book developed.

A special thank you to our "readers" Chris Anderson, Stewart Bailey, Carolyn Clarke, Niall MacKay, Eva and Elly Mackey, Mary Millen, Patti Murphy, Kathy Selecky, and to our editors Susan Kaufman, Allison Mackey, and Lesley Saringer whose perceptive comments, questions, and attention to detail were invaluable. Our thanks to family and friends who supported us on this our first major journey into the world of producing and publishing a book.

We, the authors, brought different skills and interests to the project, along with a mutual enthusiasm for researching and recording this story and bringing the past forward to the present and sharing our findings with you.

Introduction

A Twentieth Century Canadian Story

A new era was beginning in Canada in the early 1900s. The nineteenth century had been a time of pioneer settlement. The forests of southern Ontario had been cut and burned to make way for farms. The commercial lumber industry was mainly concerned with supplying squared pine timber to the British. In the mid 1800s the north was considered a "vast inaccessible wasteland" until the need for lumber developed. The Provincial Legislature decided to create a "New Ontario" in the north and, by passing the *Free Grant and Homestead Act* in 1868, encouraged homesteaders to go north and settle. The Act allowed loggers access across all land and waterways, as well as access to all white pine until the homesteaders acquired title five years after they had settled. Railways and colonization roads were started to assist the process. The financing of the exercise was the fee on prime white pine paid to the government by the lumber companies. Many of the loggers who came in the winter to cut the pine stayed and took up the free land. Many of the former textile mill workers and farmers from Europe struggling to coax a life out of the rocks of the Canadian Shield found that work in the logging camps allowed them to survive.

The great lumber baron J. R. Booth had square timber operations throughout Northern Ontario, including the future site of Fossmill. During the white pine logging era the trees were cut down and floated out to be cut into lumber in the huge water-powered mills on the Ottawa River and Georgian Bay. The lumber barons were concerned by the pace of settlement and supported the establishment of Algonquin Park in 1893, because it would prevent settlement in the vast pine forests that they wanted.

By the end of the nineteenth century most of the white pine was gone and many farms and villages had developed around the Park. With the development of steam-driven mills many entrepreneurs established local logging and lumbering operations to take advantage of the need for lumber locally and elsewhere in Ontario as well as in the United States. Eventually there were so many sawmills that loggers were running into each other in the bush and could often hear each other's mills. Large U.S. lumber operations, who had cut most of their eastern forests, looked to Canada to supply the U. S. market for lumber.

As we enter the twenty-first century we have begun to talk about building "sustainable communities" and about "sustainable development" but during most of the twentieth century the natural resources of Canada were seen as unlimited. When the natural resources were extracted from one area the companies went "over the next hill" in search of new resources to exploit. The workers in the woods and mills and their families and communities were obliged to follow.

Fossmill is one of these stories. We believe that what happened there is similar to what occurred in dozens of other operations, and in that sense, Fossmill in not a local, but a general story. The general story is made up of numerous individual stories and each of them unfolds in its own unique way. Our goal was to do an in-depth study of the life cycle of the transitory existence of a company in its various reincarnations over time, with an emphasis on the life and times of the people and personalities involved. When we started our study of Fossmill there were a few names and stories about fires, train wrecks, and other incidents, but no coherent profile of the business, the technology, or the lives of the people.

Through the sources mentioned in the acknowledgements, and with an emphasis on the interviews with former residents and/or their families, we have profiled life in a lumbering village at the edge of Algonquin Park from 1924 -1946. This core story was expanded to look at Fossmill's earlier history at Fassett, Quebec, and post history at Kiosk in Algonquin Park.

The quotations have been edited for brevity and clarity, but not in any way to change the meaning or accuracy of the statements. Space does not allow for nearly as much material as we would have liked to use from the dozens of hours of interviews.

One problem with writing books or making movies about the past is that too often the interpretation of the past is taken as the "truth" and cast in stone, and other interpretations and perspectives are lost and forgotten. Each individual involved in this story experienced Fossmill and remembers it from a unique perspective depending on many factors such as the time they were there, their social class, age, gender, and even exactly where they were at the time of individual events. We, as outsiders, also bring to the story our own biases and interests. We were delighted to find that in most cases peoples' memories were "accurate". Often we questioned the accuracy of a statement only to have it verified by other people interviewed or by documentary evidence. We have tried to reconcile the many views with each other and with the documentary evidence and to come up with a coherent story that is close to the "truth". But in the end this should more properly be called "A Fossmill Story" rather than "The Fossmill Story." In conclusion, we hope that the people of Fossmill will enjoy our reiteration and interpretation of their story and that others will gain a sense of the daily lives of those who lived it.

Doug and Paul Mackey
Chisholm Township, June 1999

Baseball: *Students play baseball in the Fossmill school yard.* [JD]

Prologue

Day of Rest

"We all beat it home."

Sunday, August 26, 1934, started out as an ordinary day off work in the lumber mill at Fossmill, Ontario. The mill workers and their families pursued their leisure time in various ways. After going to church, sport was a favourite pastime for many. The mosquito and blackfly season was finally over, and some of the workers played a game of baseball in a field west of the village.

In the 1920s, the company had cleared an area and built the ball diamond for its workers. "I played centre field," remembers Donat Huneault. "My brother Lawrence was on first base. Old man Staniforth bought us all uniforms. They were white with a black stripe. We played Chiswick, Alderdale, and Wasing—we had a pretty good team!"

"The Wasing boys and girls played the Fossmill gang," recalls Doreen Smith, who played on the team from the nearby community of Wasing. "Could Alderic Bergeron ever pitch! By the time you got the bat up, the ball was long gone."

Because of the Depression there was little money for uniforms now, but some players still wore their old uniforms with *Fassett Lumber Corporation* embroidered across the chest.

The people of Fossmill were thankful the men were still working, because the Depression had crippled the lumber industry. Many mills had closed down permanently. Their company, the Fassett Lumber Corporation, had laid off workers temporarily when the mill was closed for short periods. Fassett's timber limits were primarily in Algonquin Provincial Park, and the previous year Fassett was the only company to set up camps and cut logs in the Park. There was apprehension about the future as unsold lumber piled up in the yard.

So far, 1934 had been an eventful year: a motion picture company from England came to Fossmill to make a movie about the company; the pulp cutters in South River went on strike; a CNR train, carrying dynamite, derailed in the centre of Fossmill, killing a transient rider and almost forcing the evacuation of the village; it snowed in July, and Elzire Dionne from nearby Corbeil gave birth to quintuplets.

August was a slow month in the lumber industry. By early July the company had cut and stored its prized hardwood for drying, and the winter cutting season would not begin until September. On this Sunday, Woods Manager Jack McGibbon and his family were on vacation in Quebec, and Mill Manager Tom Howard was visiting North Bay.

Around five o'clock someone at the ball game saw black smoke rising from the direction of Fossmill. The players heard the shrill sound of the mill's steam whistle over the

crack of the bat and the shouts of encouragement from onlookers. Soon everyone was looking toward the village at a cloud of smoke billowing over the tops of the trees. Something was on fire! Was it the lumberyard, or the forest—again—or one of their houses—or even the mill?

Thirteen-year-old Margaret Gleason was watching the game that day. "Somebody happened to look up and saw all this big smoke and they said, 'Oh my God, something is burning at Fossmill.' Then they said, 'The Mill!'—we all beat it home."

They scooped up the bats, balls and gloves, and abandoned the close game. Alderic Bergeron senior's car sped past on the gravel road, a day of fishing quickly forgotten. When the villagers rounded the bend to the CNR crossing, they saw in the valley—beyond the row of houses, the school, and the church—the mill engulfed in flames. One hundred jobs in the mill, two to four hundred jobs in the bush and a decade of work building a new life and community in that isolated river valley were threatened that day.

Among the stunned crowd were people who, ten years earlier, came from Fassett, Quebec to Fossmill. The Fassett Lumber Corporation moved to Fossmill after clearcutting its entire timber limits in the mountains behind Fassett. The workers who wanted to keep their jobs were forced to move with the company.

Nora Gleason recalls the day she arrived with her family on the train from Quebec. "When we got to Fossmill we thought it was a God-Almighty place. Fassett was a good sized town. We had running water and stuff like that—things to do. There was none of that at Fossmill. There was nothing, absolutely nothing—except mosquitoes and blackflies. None of us were happy to move, but in those days a job was a job."

***Hot pond**: Steam rises from the Fassett, Quebec hot pond on a cold winter day as workers prepare to send logs up the jackladder. [UTA]*

1
The Beginning

"Fassett was a well-established town."

The Fossmill story begins in the Province of Quebec in the 1880s, when brothers Thomas and William Owens bought land at the mouth of the Salmon River (now the Riviére Kinonge) and started a white pine lumber operation. The site was a large section of the former Papineau Seigneury on the north shore of the Ottawa River between Ottawa and Montreal. The small community of Montebello, now famous for the Canadian Pacific Railway's luxury Château Montebello resort, developed there.

In 1905, the Haskell Lumber Company from Pennsylvania expanded into Canada and bought the Owens Company. It built a mill about five miles (8.5 km) east of Montebello, and a small community began to develop. Unlike many lumber operators, who leased **logging limits*** from the government, the Haskell Lumber Company owned the 70,400 acres (28 490 ha) that it logged. One of the reasons it chose the location in Quebec was the proximity of the Canadian Pacific Railway (CPR) line, which was built along the north shore of the Ottawa River in 1885.

* Use of bold text indicates that the word is included in the Glossary on page 214.

***Racicot Street**: The houses at the far end of the street are the "chemical houses" belonging to the Standard Chemical Company, the other large company operating in Fassett, Quebec.* [LD]

The Honourable Jacob Sloat Fassett (1853-1924)

J. Sloat Fassett's family were successful politicians and businessmen in the United States; Fassett, Pennsylvania is named after Sloat's father. When J. Sloat graduated from university he became a lawyer and was a District Attorney by age twenty five.

His political career began at age 30 when he became a New York State Senator, a position he held for eight years including time as the President of the Senate. For the rest of his life he was considered the leader of the Republican Party in New York State. He chaired the Republican National Convention in 1892 and was later nominated for Governor of New York State.

J. Sloat Fassett became a highly respected businessman and headed a bank in Elmira, New York as well as a ranch and cattle company. Along with his involvement in the lumber industry in Canada and the Philippines, he had gold mining interests in Korea. He died of a heart attack in 1924 in British Columbia on his way home from a trip to the Philippines. The company maintained his name at Fossmill. His obituary in the *Canada Lumberman* stated "His country has lost a patriotic public servant, the city of Elmira NY its foremost citizen and his enterprises in lumbering and other lines a great, inspiring force." A school in Elmira and a US Navy ship were also named after him.

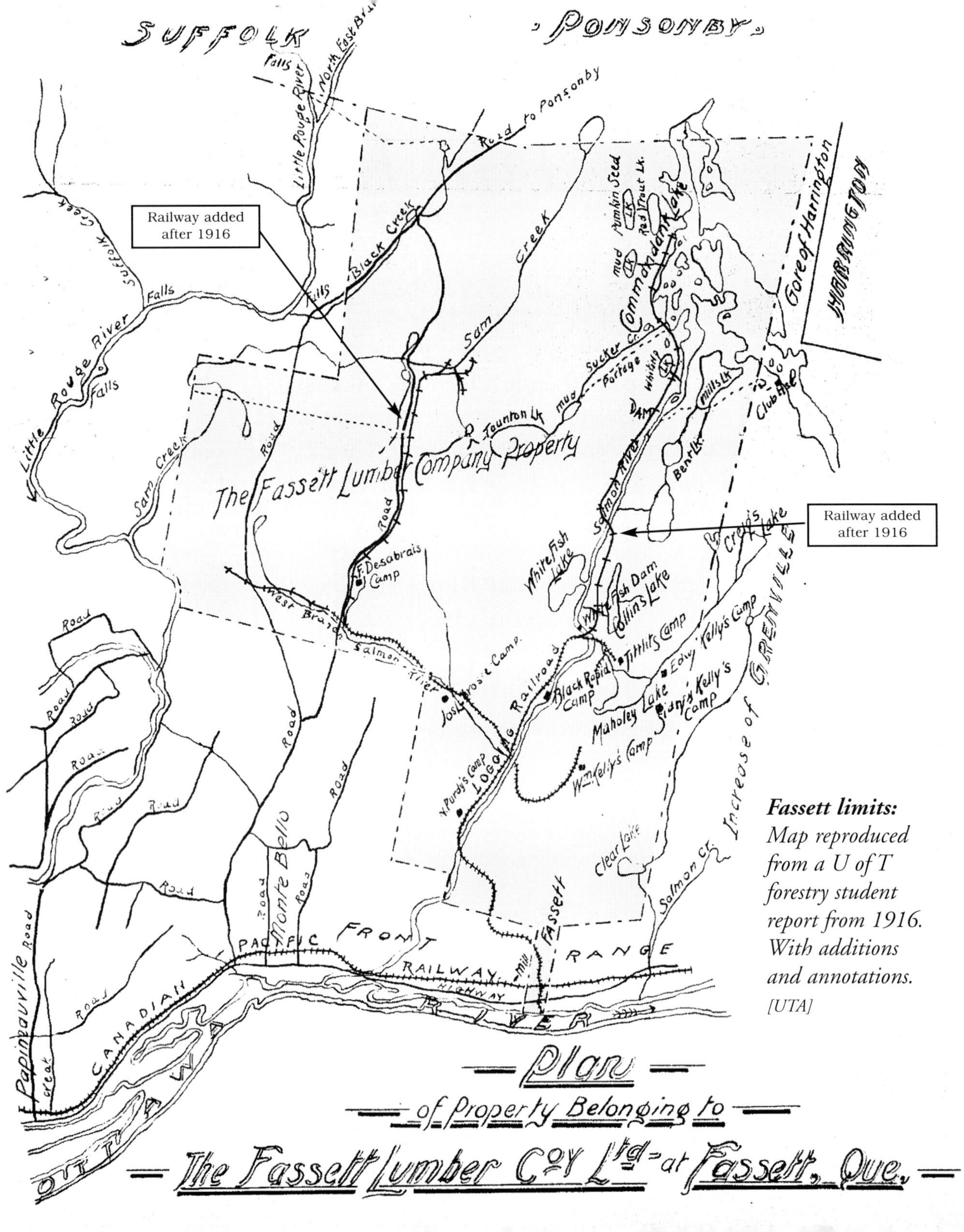

Fassett limits: *Map reproduced from a U of T forestry student report from 1916. With additions and annotations.* [UTA]

The CPR soon made the town a flag station and called it "Fassett" after the president of the Haskell Lumber Company, J. Sloat Fassett, a banker and lawyer from Elmira, New York. In 1910, Sloat Fassett bought the business and changed its name to the Fassett Lumber Company.

In the early days the manager of the company was John R. Collins, from Pennsylvania. A 1910 *Canada Lumberman* [1] article on the company mentioned that a young Canadian accountant, Sidney J. Staniforth, was working for the company.

Logging was in Sydney Staniforth's blood. William Staniforth, his father, built a sawmill in the Lachute area of Quebec in 1883. William logged in the Mont Tremblant area, floating his logs down the Rouge River to the Ottawa River. Today, Staniforth Lake, which flows into Lac Tremblant is named after him.

Sydney was born in 1882. He attended the Lachute Academy and business school in Montreal. In 1908, he started working as an accountant for what became the Fassett Lumber Company. He did not stay an accountant for long. By 1913 he was taking over management of the company when Collins was on vacation, and soon replaced Collins full time.

Office staff: *This photo from the* Canada Lumberman *shows from left to right the salesman, secretary, and accountant S. J. Staniforth. [CL/UTL]*

Sydney Staniforth: *(Above centre) Shown here as the manager of a Fassett Lumber Company sponsored baseball team in 1910. [RS] (Below) This portrait of Sydney Staniforth appeared in many issues of the* Canada Lumberman *in the 1920s and early 30s. [CL/RS]*

Building Fassett

On the north side of the CPR tracks, the company built its office, a lumber mill, an engine and boiler house, a planing and shingle mill, dug a log pond, and created a lumberyard. They built several railway sidings connecting to the CPR main line. A spur line extended across the CPR to a company dock on the Ottawa River where lumber was loaded

Fassett Lumber Company office: *the logging limits are in the background, and the CPR passes in front of the office. [DS]*

onto barges for shipment to Montreal and the eastern United States.

South of the CPR, on flat land between the tracks and the Ottawa River, the company established the town of Fassett. Some streets became "company streets" where the tenant-residents were exclusively employees of the Fassett Lumber Company. Fassett, however, was not a company town as many resource industry towns were, or as Fossmill was later to become.

A 1906 *Canada Lumberman* article reported that "The town has grown rather rapidly and boasts one large general store, three grocery stores, two meat markets, two hotels, two blacksmith shops, a school house, a church, and fifty houses." A new post office was established, and the CPR built a station including telegraph and telephone equipment. A turbine pumped water from the Ottawa River to the millpond and provided fire protection for the town and the company.

A second major industry, the Standard Chemical Company, built a plant next to the Fassett mill around 1912. It utilized the small hardwood trees that were not taken for lumber by Fassett. These trees were baked in special ovens to produce charcoal and various chemical by-products (wood alcohol, wood oil, acidic acid, formaldehyde, acetate of lime). Many of the town's residents were **jobbers**

Fassett, Quebec's french school: *A class of girls poses in front of their school. [LN]*

Pennsylvania Street: *The English school/church is the building on the right. [DS]*

who were independent contractors working for the two main industries.

Fassett was largely a French-Catholic community with a French school and a large stone Catholic church. However, there was a small English community, which included the managers of the main industries, who wielded considerable influence in the town. Most of the English lived on the same street. June Summers (Maxfield)*, the daughter of the Fassett paymaster Roy Summers, remembers going to the English school, which was in a small building on the same street where she lived. Protestants held church services in the same school. "We were in a minority, a very small English group," Summers notes.

In 1918, Fassett became a town with a municipal council, and by 1921 it had a population of 1,512.

The Salmon River & Northern Railway

The Haskell/Fassett Lumber Company established the Salmon River & Northern Railway (SRNR) to haul its logs from the bush. The SRNR was a major factor in the company's success. Its 28 miles (47 km) of track, in two branches, penetrated deeply into the 110 square mile (284 sq. km) Fassett property. Many temporary switches were built and relocated, allowing Fassett to haul out previously inaccessible hardwoods.

* In this book we have chosen to refer to women by the names they used at the time of the events being described, even though many used the surname of their husbands at the time of the interviews. The first time their name is mentioned we have included their married name in brackets.

Town store: *Fassett residents posing outside a local grocery store. [LN]*

Sunday best: *Fassett children out for a a stroll. [DS]*

Locomotive No 51: *Fassett's first Shay locomotive in front of the company office. Morris Donahue is at right standing on the locomotive and manager Sidney Staniforth is the man with his hands in his pockets.* [ON]

"Cutting hardwood was something new," said Stella McGibbon, wife of Fassett's **woods superintendent** Jack McGibbon. "In our part of the country, Fassett was the first to touch hardwood. They brought it in by rail because you can't float hardwood. The idea, to bring logs in by railroad, came from the States." Using a railway allowed the company to add hardwood to its product line. Hardwood eventually became the main focus of its operation in Fossmill.

The company's railway rolling stock consisted of: three **Shay-geared locomotives**, No. 51 (50 tons), No. 53 (60 tons), No. 54 (70 tons); one **saddle tank switching engine**, No. 52 (25 tons); two **Barnhart steam log loaders**; 60 log cars; 25 flat cars; a snowplow, and a Model-T **railcar**. One Shay remained in

Story continued on page 25 *

* At various places in this story the narrative is interrupted by a few pages of photographs or major background sidebars with a grey background, before continuing.

Tank locomotive No 52: *This little engine worked around the mills' lumberyard.* [UTA]

Railcar: *This Model T had steel flanged rings fitted over the rubber tires to allow it to travel on the company railway.* [UTA]

Leading edge technology: *An ad for Shay locomotives in 1906. American inventor Ephraim Shay designed the Shay-geared locomotive in 1880 specially for logging and mining railways. Vertically mounted pistons turned a system of rods, slip-joints, and gears that were connected to and drove all the wheels on the locomotive. This unique design allowed the locomotive's wheels to adjust to the bumpy and uneven track inevitable in logging operations. A Shay locomotive could pull heavy loads up 10 to 14 percent grades, much like a four-wheel-drive truck in first gear.(See appendix A for further details).* [CL/RBL]

A Ride on the Salmon River & Northern Railway

The following is an excerpt from a report on the Fassett operation written by A.W. Bentley and T.A. Clarke, students in the University of Toronto Forestry Department in 1919. In this excerpt they describe the experience of a trip on the logging railway.

The track follows the bank of the Salmon River and the West Branch crossing from side to side as the banks afford sufficient ground for the road bed. The bridges crossing the river are of the wooden trestle type of construction supported on piers and those that have been built for some years are somewhat flexible. One span, about 200 feet long and some forty feet above the stream, if crossed after sundown presents no fears but only a delightful swinging sensation. The road is really a succession of alternate curves religiously following the bank of the stream. Guard rails were noticed on a few of the real sharp curves where two feet from the outside rail was a clear drop to the water below.

Like all railways minor accidents do occur. On our trip down a loaded log car broke an axle and tore up a few ties before the train could be stopped. The logs were unloaded beside the track, the flat car was removed and the journey continued. A few miles further on, the middle trucks of the locomotive jumped the track on a curve and the train was rudely brought to an abrupt stop with the locomotive standing with but a comfortable six inches to spare from plunging down the bank of the river below. Two of the driving rods were found to be broken and an effort was made to get the locomotive on to the rails again with the remaining driving rod. The attempt was unsuccessful and the rod was broken during the struggle. This left the train about six miles from Fassett. Some of the crew set out for town and another locomotive was obtained and the train brought in about four o'clock the following morning. This experience is not uncommon especially in winter time.

Most of the equipment has been bumping along over this road for some ten or twelve years now and is not what it once was. The usual speed on the road is between six and ten miles per hour. In a general sense the grade is up hill from the mill at Fassett to Lake Commandant. This is not a serious drawback and is compensated for by the ease with which the loaded cars of logs are brought down to the mill. The train can "coast" most of the way from the woods to the mill.

Salmon River & Northern Railway: *Shay locomotive and log train returning to the mill from the limits in the mountains.[CL/RS]*

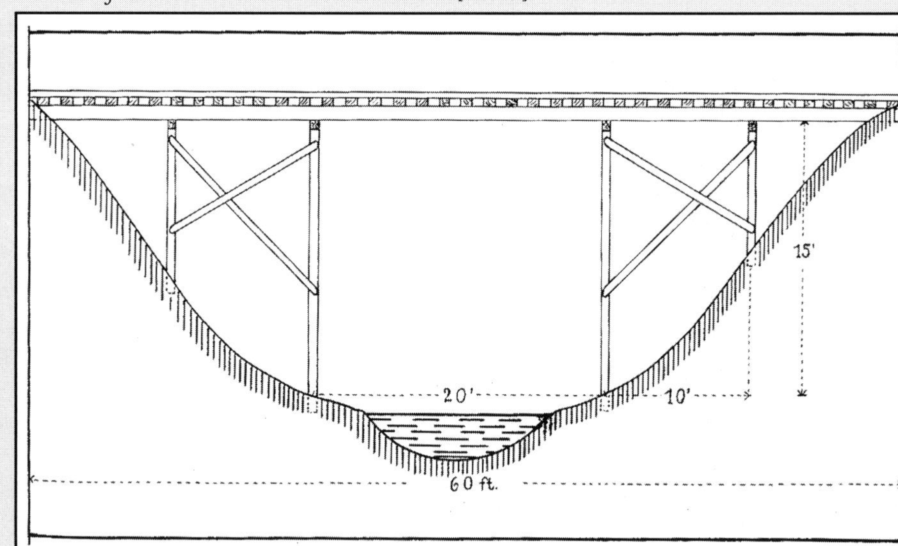

Salmon River & Northern Railway bridge: *Diagram from a 1916 forestry student report on the Fassett Lumber Company. This one is only 60 Feet (20 m) long.*

Railbed: *(Above) The Salmon River & Northern Railway winds through the challenging terrain of the Fassett limits.* [UTA]

Frosty logs: *(Top left) Logs recently delivered from the bush have been unloaded and will soon be thawed out in the mill's hot pond.* [UTA]

Fassett engine house: *(Left) Locomotive No. 52 and a Barnhart loader of the Salmon River & Northern Railway.* [RF]

***Abandoned**: After removing the trees from the area this camp by the railway was abandoned and the lumberjacks moved to a new site on the Fassett limits.* [DS]

Story continued from Page 21

the woods to work with the two loaders, while the other two hauled the logs. The saddle tank engine worked around the mill, and the railcar was used for supervision and inspection.

To avoid the expense and inconvenience of operating **logging camps**, the company hired jobbers to carry out the cutting, skidding, and hauling of logs to the railway. The jobbers also cut and hauled huge quantities of **tanbark**, **pulpwood**, and **cordwood** for the company.

Fassett, like most logging companies at the time, clearcut its limits with little thought for the future. The *Canada Lumberman* stated in 1913 that Fassett's limits would be "completely cut over before operations upon it come to an end. Even the small stuff will be taken, a market for it being found with the chemical company operating at Fassett." In 1920, in spite of rugged terrain, steep grades, long distances, and devastating weather, the railway transported 30 million **board feet** of logs, 20,000 **cords** of firewood, plus tanbark and pulpwood to the mill.

The sawmill

Haskell had installed used equipment from Pennsylvania in the original mill. After a fire destroyed the mill in 1913, the company (now Fassett) installed the latest equipment in a new mill. The new mill could produce 100,000 board feet of lumber in each ten-hour shift. Eight loading docks handled 25 million board feet of lumber a year from the new mill. Workers placed the lumber on flatbed railcars called **trams** and pushed them along narrow-gauge track mounted on elevated **tramways,** to

Winter camp: *This camp, built on the mountainside, was home to around one hundred lumberjacks during the winter cutting season. All the stoves are working full blast to fight the cold.* [UTA]

Horse power: *Dozens of teamsters and horses were used to haul the logs from the cutting site to the railway.* [UTA]

Tramways: *Sawn lumber was pushed on these tramways to the lumber piles in the yard. The Standard Chemical Company plant is in the background.* [CL/UTL]

Mill from the mountain: *An early photograph of the Fassett mill. The Ottawa River and town are in the background. The Standard Chemical Company has not yet been built beside the mill.* [NG]

the yard for drying. Once dried and sold, the lumber was loaded into boxcars and hauled to the CPR, or it was loaded onto flatcars and hauled to the company dock on the Ottawa River, for shipment. A conveyor transferred **slabs** from the Fassett mill east to the nearby Standard Chemical Company mill, for use in firing its ovens.

Accident

Several people including Sydney Staniforth's son, Donald, remember a dramatic incident involving the railway siding that ran south down the middle of a main street, passing houses and stores to the company dock. Donald says that one day the train was "going down to load timber on the barges. The engine was behind and the loader and lumber cars were up front and they uncoupled from the engine. Once that happened, the whistle blew and it was downhill all the way to the dock. The wood and loader went right off the end." Fassett

Standard Chemical plant: *Distinctive triple stacks of the Standard Chemical Company plant.* [UTA]

River transportation: *Barges were used to ship Fassett's lumber on the Ottawa River. A Barnhart loader and lumber cars are on the pier.* [DS]

Valuable salvage: *Loader and log cars are recovered from the bottom of the river.* [DS]

resident Louis Demers recalls that a barge had just moved away from the end of the dock and that the loader and lumber cars went tumbling into the Ottawa river.

Tourists: *A group of visitors (bottom right) inspect the loading operation at a log dump on the Fassett limits. Louis Demers recalls that during one such visit two women were up in a loader "when it broke loose from the engine and rolled away with no brakes."* [ON]

All dressed up: *Young boys on the Fassett, Quebec boardwalk.* [LD]

Fassett fun

The Ottawa River to the south and the Laurentian Mountains to the north are the focus of many Fassett memories. Donat Huneault remembers that his "oldest brother, Jack, worked on the steam loader. Everything came down on the railroad from way back. My Dad and I used to go fishing up there on the Salmon River."

"That was quite a mountain." Nora Gleason (Clark) recalls times she spent with her friend Genevieve Summers on one of the hills behind Fassett. "I can remember a cave Genevieve and I had up there. We were only little kids—today, I wouldn't want my kids to go that far from home."

June Summers remembers breaking rules against going down to the Ottawa River and the company dock. "We were warned not to go near the river unless there were adults with us, but we used to sneak away and go anyway." She also recalls some parent-supervised activities. "There was the old swimming hole on the point and we all went there to swim. I remember Dad pushing us off the pier there, to make us 'sink or swim'. Dad had a little red launch and every year we went up the Ottawa River on the Ontario side and camped. I remember the whole family looked forward to that. It was beautiful there in the hills, especially in the fall."

Fassett, Quebec 1928: *This early aerial photograph was taken shortly after the Fassett Lumber Company left for Ontario. The Standard Chemical Company was still operating at the time.* *[NAPL #HA523-52]*

Sunday best: *A Fassett family on the company's dock on the Ottawa River.* [DS]

Winter's cut: *Loader and logging train loads logs at a log dump ready for transportation to the mill.* [CL/UTL]

Ontario bound

The Fassett policy of **clearcutting** the forest meant that eventually the company would have to move on. The *Canada Lumberman* reported in 1920 that the company had enough trees for four more years. By 1924 Fassett had cut all of its trees except the smaller ones that the Standard Chemical Company wanted. In 1922 the Fassett company had begun looking for new limits and had sent its woods superintendent to Ontario to look for a choice location.

Woods Superintendent John "Jack" McGibbon played a major role in the Fassett Lumber Company's success. He **walked** or **cruised** the bush to determine the amount of timber available, decided where cutting would take place, laid out the roads, picked the campsites, hired the jobbers, kept supplies in stock, and, along with his foremen, kept the logs moving to the mill. Jack was born in Lachute, Quebec and was the eldest of five children of A. D. McGibbon, a successful lumberman from a family that started in the business in the 1850s. Jack joined the expanding Fassett Lumber Company in 1916 and was soon responsible for the complete logging operation.

By 1922, Fassett specialized in hardwoods and was interested in the rich hardwood stands on the western side of Ontario's Algonquin Provincial Park. Other companies were using railways to haul logs from this area. The Standard Chemical Company had a logging railway and mill, cutting and hauling logs from limits east of South River (see sidebar, page 34). In the early part of the century, the Trout Creek Logging Company operated a logging railway from Trout Creek into the west side of Algonquin Park. One of the main players in this company was William Foster.

William Foster

Foster was originally from Hawkstone in Southern Ontario, but by as early as 1896 he had acquired logging rights to the south end of Chisholm Township. He built several logging camps and mill sites in various places in south Chisholm, hired many local farmers, and used oxen as well as horses to haul his logs. In 1908, in alliance with other lumbermen in the area he formed the Ballantyne Lumber Company, which logged in Chisholm and in Ballantyne Township to the south. In 1911, the Ballantyne Lumber Company amalgamated with other companies in Trout Creek to form the Dominion Wood and Lumber Company. This company formed the Trout Creek Logging Company Limited in June 1911 specifically to build the Trout Creek Logging Railway east from Trout Creek. William Foster was a director of the company and was responsible for building the railway (see sidebar, page 33).

[2]

For the Fassett Lumber Company the problem with the area west of Algonquin Park was that the transportation routes from the mills ran south to Toronto rather than east to Quebec and the traditional Fassett buyers. Fassett needed a source of hardwood that could be shipped easily to Montreal and the eastern United States.

Story continued on page 41

Oxen: *William Foster's lumberjacks used oxen and horses to haul logs from the bush. Oxen "could go through muskeg like a moose."* [CWI]

Foster's hunt camp: *William Foster is the man in the centre with the boy on his left.* [CWI]

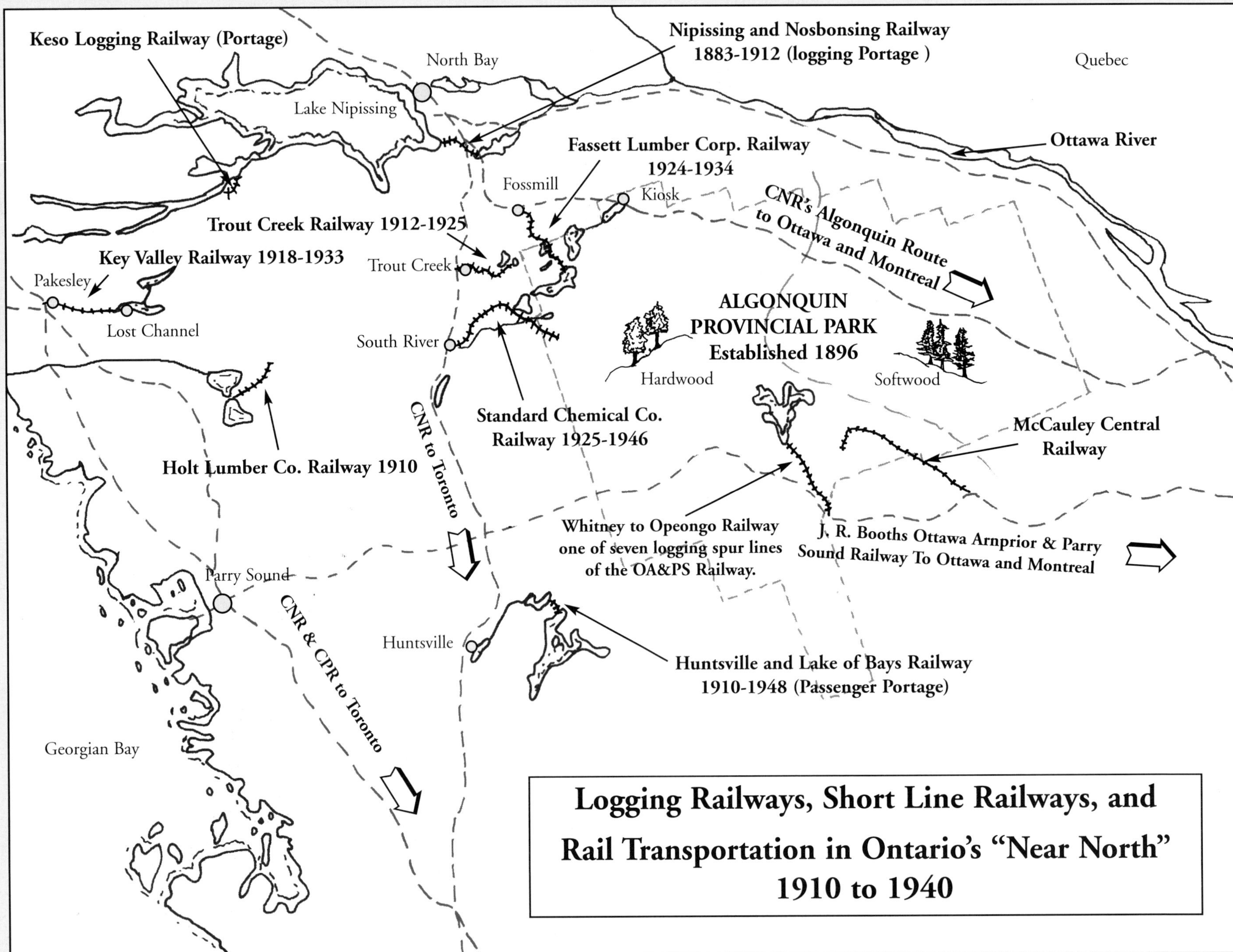

Logging Railways, Short Line Railways, and Rail Transportation in Ontario's "Near North" 1910 to 1940

[JT]

Trout Creek Logging Railway

The first sawmill at Trout Creek was built in 1889 beside a waterfall on Trout Creek. It used a water wheel for power. The original mill burned in 1892 and in 1899 the land was acquired by brothers Gilbert and James Trussler. The mill was rebuilt and by 1910 Trussler Brothers Limited was the largest employer in the area. The Trusslers blasted a wheel pit into the rocks beside the creek and added a modern turbine to drive the mill.

Also operating in the Trout Creek area were the Kaufman Lumber Company and the Ballantyne Lumber Company. The Trussler Brothers, Kaufman, and Ballantyne companies amalgamated in 1911 to form the Dominion Wood and Lumber Company (DWL). They operated just east of the Grand Trunk Railway (now the CNR) on the edge of Trout Creek. They also built a chemical plant to make charcoal and distillates from the smaller cordwood left after the main logging was completed.

The Trusslers built their own general store for the community. The mill workers were paid with company script and had to spend it at the store.

In 1917, because the rivers east of Trout Creek were not big enough to float logs and some of the timber was too far away for teams of horses, the DWL built its own railway and formed a separate railway company, the Trout Creek Logging Company Limited, to access and carry its logs. [3]

William J. Foster, one of the owners of the Ballantyne Lumber Company, played a key role in the DWL Company and in the development of the railway. Hartley Trussler, one of James Trussler's sons, worked part-time for the DWL and later became a prolific local historian and an excellent source of information on those early days. In a 1974 column in the *North Bay Nugget* he wrote that the Trout Creek railway went 15 miles (24 km) into Algonquin Park and that "W.J. Foster was something of an engineer, improviser and contractor. There was no survey made or specifications. He used his experienced eye to judge the grades. Curves and tangents were of little importance. Bulldozers and motor trucks were unknown at that time so the line was built by man and horse power with axes, picks and shovels and two handed horse drawn scrapers. It was amazing how things were accomplished in those days." He wrote that the right of way followed the contours of the land and high spots were leveled and low spots filled. Bridges and culverts were made of timber cut along the way. One case of dynamite was used to blast a ledge of rock on Buck Lake. The next year a steam shovel was purchased, ballast cars built and the rail bed was firmed up.

A locomotive was purchased new from the Heisler Locomotive Works in Erie, Pennsylvania. It was a 26 ton 2 truck engine (serial #1258) and was delivered in 1912. In 1917 they purchased a used 2 truck shay locomotive from a defunct logging railroad in Marquette, Michigan. Hartley wrote, "Both the Heisler and the Shay were geared locomotives and when you were going 15 mph the roar of the exhaust sounded as if you were going 60 mph. The Heisler was a well built efficient machine but the old Shay was a rattletrap. When it was doing 15 mph you would think that it would shake itself to pieces."

The Trout Creek Logging Company was dissolved in 1925 and its assets transferred to the parent Dominion Wood and Lumber Company. The Dominion Wood and Lumber Company continued until 1932.

In 1928, the Heisler engine was sold to the Bethune Lumber Company in Huntsville where it had a few years of limited service. It was eventually scrapped in the early 1950s, after sitting in storage for several years. The old railway bed now forms part of the Ministry of Natural Resources forest access road east of Trout Creek. This road joins a road that was once the rail bed for the Standard Chemical Company railroad and ends at South River.

The Shay [KC]

The Climax [JM]

South River Logging Railway

There were several logging companies in the South River area but the largest was the lumber division of the Standard Chemical Company. Its lumber division had upwards of 165 employees in South River and 150 to 350 in the bush in season.

The Standard Chemical Company plant in South River was one of ten plants in Ontario and Quebec producing charcoal and distillates. From its start in 1904 it employed about 100 people until it became dormant after World War One. In order to access the hardwood east of South River the company built a 1000 foot (305 m) trestle bridge across the South River and originally built a railway 4 miles (6.5 km) into the bush. When it started operating again in 1925, it decided to add a lumber division and mill and expand the railway.

The railway eventually went 25 miles (40 km) east to Winifred Lake in Algonquin Park. A. F. Cooper, the woods manager, brought in his son Orlo to build the railway and run the operation. They used a 50 ton Climax-geared locomotive and an 80 ton Shay-geared locomotive to haul the log cars. A Barnhart log loader was used to load logs onto the cars.

During World War Two, when not enough local men were available, Mr. Cooper arranged for a couple of hundred German prisoners of war to work in the bush.

By the 1940s the railway was replaced by trucks and the rails were pulled up. The Shay locomotive was sold to the Abitibi Paper Company in Iroquois Falls, where it remains today as a tourist attraction (see places to visit page 222). Orlo Cooper, who replaced his father as head of lumber division, sold the company in 1951, and the mill later burned down.

The following quote from a forestry student report describes the railway:

> The construction of the railroad is very primitive and far below the standard of ordinary railroads. Many steep grades are encountered and the ties are often made out of the trees standing on the right of way, thus ties of green hardwood are seen along with ties cut out of dead black spruce. Ties from side switches are taken up after the switch has ceased to be used and are used on further extensions of the line, therefore, the movement of the cars is very irregular and frequently far from being smooth. In one particular place, near the old camp of last winter, derailing takes place quite regularly, this is where the extension joins the old line; however, numerous levers, stakes and iron bars indicate this place and everybody is prepared for the accident. When the train approaches the old camp, the engineer slows down the speed and carefully goes over the spot, some of the cars derail and are set in place after some strenuous work and the train is then ready to proceed further. Water is also taken in by the engine in places where the railroad crosses open creeks and the train has to stop before some steep hills in order to get enough steam and impetus for climbing them.

Early Logging and Lumbering in Chisholm Township

There is a long history of logging and lumbering in Chisholm Township and in the surrounding area. J. R. Booth's lumberjacks came to Chisholm for squared timber in the 1860s. Booth's agent, David Moore, logged in Calvin Township (Moore Lake) and on Lake Talon on the Mattawa river. Booth came to Chisholm in 1860 and set up a depot east of Wasi Lake on the Nosbonsing River, three kilometers south of Lake Nosbonsing. The Nosbonsing river flowed from deep in Bonfield Township to Lake Nosbonsing, providing transportation for thousands of squared timbers and pine **saw logs**. The logs were towed across the Lake and down the Kaibuskong River to the Mattawa River and down the Ottawa River to Hull, Quebec and Booth's mills.

Chisholm's geography was a challenge to Booth's operation. While the waters of the

Alligator: *Home-made tugboat based on the Alligator steam warping tug tows Booth's logs on Wasi Lake.* [CWI]

Local mill: *(Left) The Lewis and Spears mill on the South Himsworth boundary with Chisholm.* [CWI]

First loggers: *(Below) The J. R. Booth Depot and farm, the headquarters of Booth's operation in Chisholm Township.*[CWI]

Nosbonsing river eventually emptied into the Ottawa River, the adjacent waters of the Wasi, Graham, and Genesee creeks emptied into Lake Nipissing and Georgian Bay, in the opposite direction from Booth's mills on the Ottawa River. In Chisholm he built a dam at the outlet of Wasi Lake, flooded a large part of Chisholm, and proposed to build a canal to the Nosbonsing River. The canal was never dug and the logs were floated to Lake Nipissing instead. Booth eventually built one of the first logging railways in Canada, the Nipissing and Nosbonsing Railway, to carry his logs from Lake Nipissing to Lake Nosbonsing. He eventually logged most of Chisholm and area and acquired limits in Algonquin Park.

Many of Chisholm's early settlers came as lumberjacks for J. R. Booth and other early settlers worked in the lumber camps in the winter. Well known Chisholm lumberman Joe Chayer senior logged Chisholm for Booth in the early days.

With the development of a permanent local community, many entrepreneurs took advantage of the hardwoods and pine left by Booth, and established mills in Chisholm (see Chisholm Map page 53). They hauled the finished lumber by sleigh, west to Powassan and the Grand Trunk Railway. One was Robert Thompson who came to Chisholm in the 1890s and built a depot and small community in the southwest corner of Chisholm. Bill Foster bought this property in 1902 and began many years of logging in Chisholm. His fourth location was on the newly built CNR where he established "Foster's Mill", or as it was sometimes called, "Fossmill".

Bucking: *Local Chisholm lumberjacks cut a tree into log lengths. [CWI]*

Local lumber: *Lumber from the small mills operating in Chisholm was hauled to Powassan by sleigh in the winter. [JE]*

Firewood: *Slabs from the mill are piled on a sleigh to be used as firewood. [CWI]*

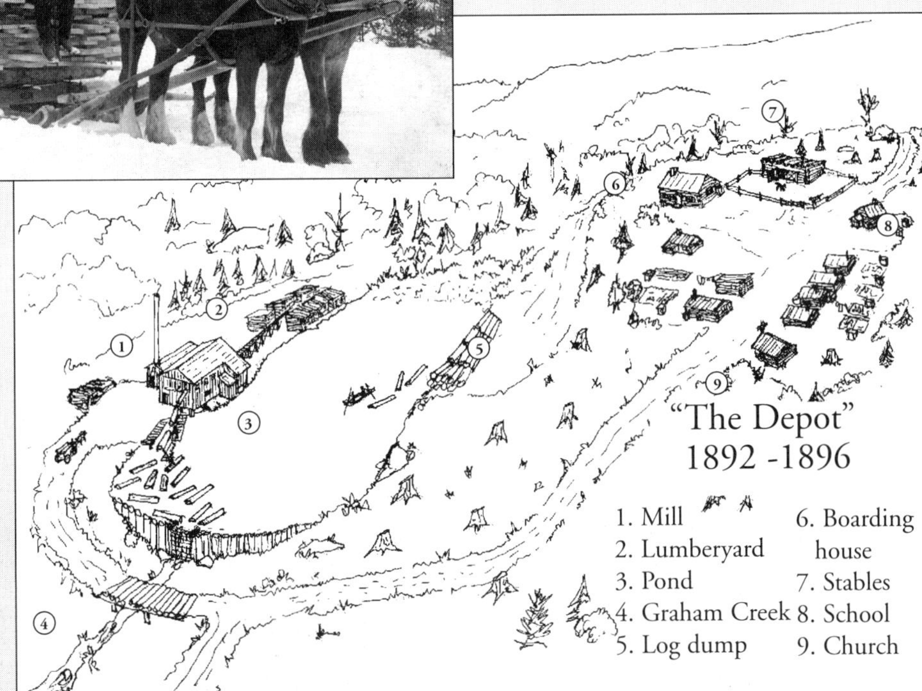

Construction train, 1913: *Hundreds of immigrants hired to build the line lived in local houses and in tents nearby. [CWI]*

The Canadian National Railway's Algonquin Route in Chisholm Township

In the 1880s, when the transcontinental Canadian Pacific Railway was built, it bypassed Algonquin Park for a more northerly route. When a second transcontinental route was built in 1915 the Canadian Northern Ontario Railway went through 130 kilometers of the Park and went north-west from there to North Bay, bisecting Chisholm Township.

The planning, construction and eventual constant movement of trains changed the Township in many ways, both positive and negative. One change was the coming of lumbermen attracted by the improved transportation. One of them was William Foster, who built his mill between the Wasi river and the railway, and another was J.R. Booth, who used sidings at Fossmill.

The railway went bankrupt in 1918, and was purchased by the Canadian government. In 1923 it was amalgamated with the Grand Trunk Railway to form the Canadian National Railway (CNR).

The railway continued to operate into the 1990s when its cost of operation became an issue. The CNR was a 95 million dollar drain on the federal coffers over the years and was finally privatized.

Alderdale trestle: *After the line was in operation tons of fill and ballast were dumped into the spaces around the wooden frame eventually covering it completely and forming an embankment.* *(see photo of Alderdale train wreck page 181)* *[CWI]*

Trestle building: *To keep the line as level as possible wooden trestles were built to bridge low spots and valleys. [CWI]*

Removing the line: *In 1996 the rails were removed from th line. Here they are pulled up from the Alderdale trestle. [DM]*

Making a cut: *A steam shovel digs a path for the line through a hill. A work train on temporary track hauls away the excess to be used as fill in low areas nearby. [DA]*

CNR stock went on sale on Canadian and American stock markets. As part of a major cost-saving effort the Algonquin route was abandoned. On November 24, 1995, almost 80 years to the day after the railway opened, the last train went through. In short order the rails and ties were removed, leaving the railbed as the only tangible evidence of the line.

Efforts to develop the line into a trail system met with some success to the west between Sudbury and North Bay and on the east side of the Park. An effort is being made to develop a trail with an environmental, educational and heritage theme in the Park from Kiosk to Brent. Perhaps some of the excitement of the old logging days will be revived and maintained for the future.

Local centre: *Alderdale became a local shopping spot with the establishment of grocery stores. [CWI]*

Work crew: *Many Chisholm people were hired to work on the CN Railway. [CWI]*

First train: *In November 1915 the first train was sent through Chisholm and a new era had arrived. [CWI]*

Dismantling Alderdale station: *When the railway was rumoured to be built, some people bought land around Alderdale thinking it would be a divisional point, but it was never more than a watering station, and with the arrival of diesel locomotives, it lost even that status. [CWI]*

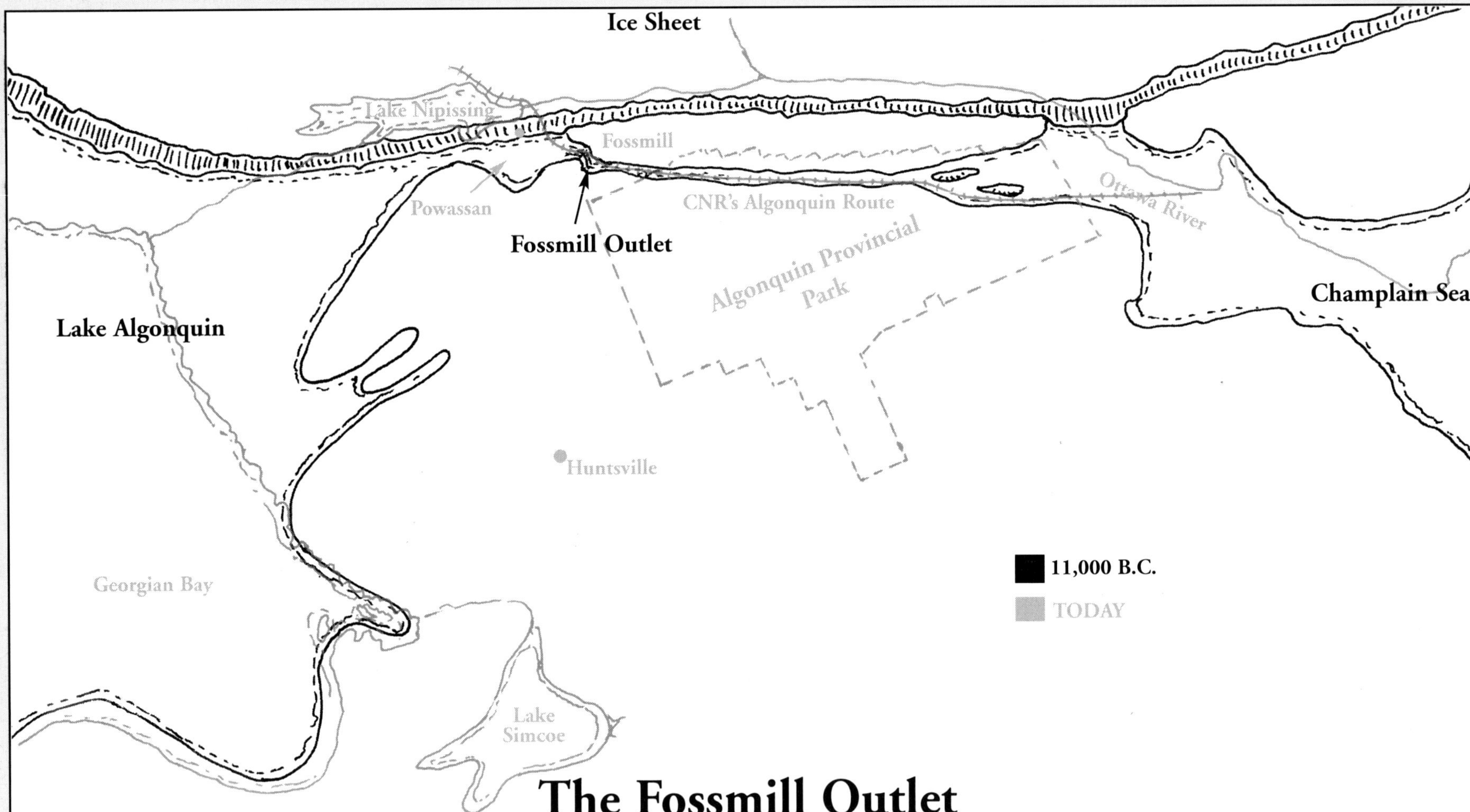

The Fossmill Outlet

Fossmill's place in history goes back 13,000 years to the end of the last ice age. As the ice sheet receded at a rate of approximately 134 meters per year a huge lake called the "Champlain Sea" by geologists filled the Ottawa valley and the St. Lawrence. Another huge lake called "Lake Algonquin" covered the area west of Fossmill including all of Georgian Bay. Lake Algonquin was 50 to 100 meters higher than the Champlain Sea. The water from Lake Algonquin eventually overflowed its shores in the area of present-day Fossmill, creating a huge waterfall several times the size of Niagara Falls. For several centuries a river the size of the St. Lawrence River cut across eastern Ontario eventually becoming the Petawawa River. The valley created by this River became the path of the Canadian National Railway's Algonquin route. Geologists have named the breakthrough point of this monumental event the "Fossmill Outlet" and Fossmill has found a place in Canada's geological history.

Untouched forest: *An example of the mixed hard wood and softwood that the Fassett Lumber Company wanted.* [UTA]

Story continued from Page 31

By 1915, the Canadian Northern Ontario Railway, which in 1918 became part of the Canadian National Railway (CNR), had built its railway through Algonquin Park to the Ottawa River, with connections to Montreal. The line passed close to the hardwood stands in the north-west corner of the Park. In 1915, seeing the advantage of this transportation route, William Foster built a small mill between the CNR and the Wasi River in the south-west corner of Chisholm Township on the edge of the Park.

"Foster's Mill"

Foster's Wasi River operation had a 6.5 square mile (17 sq. km) cutting limit, a small

"Algonquin Park Lumber Saved from Destruction"

An article in the *Canada Lumberman* in 1910 praised the government for buying the limits of a logging company in Algonquin Park and prematurely stated. "Now that lumbering on an extensive scale has been stopped in the park, there is no intention on the part of the Government to allow the timber which grows to maturity to decay and become of no value. A scientific system of forestry is contemplated, under which the forest growth will be thinned from time to time to permit the younger trees to grow to the most advantage."

Foster's mill: *William Foster is standing to the left of his crew.* [CWI]

mill, office, blacksmith shop, some houses, a cookhouse, and a boarding house, which was run by a woman named Mrs. Short. Foster loaded his sawn lumber into boxcars on a railway siding he built beside the CNR. J. R. Booth, Canada's most successful lumber baron, had a siding nearby where his men loaded logs from his Kiosk limits to the east.

Foster shipped lumber, ties, and logs from his mill. Chisholm Township's local history book, *Our Chisholm Story* [4], states that it was because these logs were "spotted at the eastern seaboard en route for overseas markets that inspired the Fassett Lumber Co. to send a capable man of John McGibbon's caliber to travel the limit and learn its true value. With a local guide he did so and the limit proved satisfactory."

Jack's daughter, Margaret "Peggy" McGibbon (Brushey), remembers that her father "came up here in 1922 and '23 and cruised the limits thoroughly for miles and miles. He said there was at least 15 years work."

What most interested Jack McGibbon was the limits of the Booth Lumber Company just three miles (5 km) south of Foster's mill in the highlands of Algonquin Provincial Park. Booth had cut most of the white pine from the area, but it was still covered in untouched stands of mixed hardwood as well as some prized white pine, which Booth had left because it was too isolated to take out profitably. On Jack's recommendation the Fassett Lumber Company negotiated the lease of the cutting rights to Booth's dormant limits in the Park, and decided to buy Foster's property and relocate its mill to Chisholm Township.

Fassett Lumber Company President J. Sloat Fassett died suddenly while on vacation in

***Foster's mill:** Lumber is pushed on a tram from the mill to sidings by the CNR.* [CWI]

1924. Canadian investors bought the company, reorganized it, changed the name to the Fassett Lumber *Corporation*, and moved the offices to Montreal. Charles Dale became president of the new half-million-dollar corporation, Sydney Staniforth became vice-president and managing director, and Harold Skelton, president of Fassett's accounting firm, became secretary-treasurer.

The company's workers had made a comfortable life in Fassett, Quebec, but this was to end as they reluctantly relocated to the hamlet that was variously called Foster's Mill, Foss Mills, Fosmill and eventually Fossmill in the isolated Wasi River valley on the north-west edge of Algonquin Provincial Park. "They were pretty loyal," recalls Donald Staniforth. "They must have been loyal, because they went from Fassett to Fossmill. But I suppose in those days things were so tough, it was the only place they could get a job. It was pretty rugged in Fossmill. Fassett was a well-established town."

The company sold Shay locomotive No. 53 to the Standard Chemical Company and maintained an office in Fassett to sell the lumber remaining in the yard. The rest of the equipment was relocated to the new operation, which was under construction at Fossmill, and a new era in the company's history began.

Fassett Quartet: *A meeting of the early Fossmill development team. Tom Howard, Ross Todd, Sydney Staniforth, and Jack McGibbon.* [DS]

Early Fassett office: *The original development decisions were made at meetings in this building.* [DS]

A new mill: *Scaffolding surrounds the incomplete smokestack of the new Fossmill mill in 1924.* [DS]

2

A New Home

A New Home

"Nothing except mosquitoes and blackflies."

He wouldn't stop barking. Teddy was a small collie that belonged to John Gleason who was seven years old when he moved from Fassett to Fossmill. "That dog wouldn't stay in Fassett when I was leaving," John remembers. "We built a crate for it so we could bring it on the train. The dog had to go in the baggage car because they wouldn't let us take him in the passenger car. When the train got out a ways, a fella from the baggage car came in and asked my Dad if I could go back to see the dog because the dog was going crazy back there. Well the dog quieted down as soon as he saw me. The baggage man said, 'It's against the rules, but maybe you better take the dog up the train.' The dog slept up on the seat with me for the rest of the trip."

John and Teddy sat with John's father, Allan Gleason. As there were few houses in Fossmill, Allan was going ahead to work and to build a house for the family. He was

Winter 1924: *Cold has driven everyone indoors by the fire. All of the buildings seen here, except the new store and boarding house, were Fosters and were eventually torn down to make way for new ones.* [DS]

***Gleason family**: (Top) Allan and Grace Gleason, Grandma Gaston, and Lester and Nora Gleason. John Gleason (left) and Margaret Gleason (right) were born later.* [MF&NG]

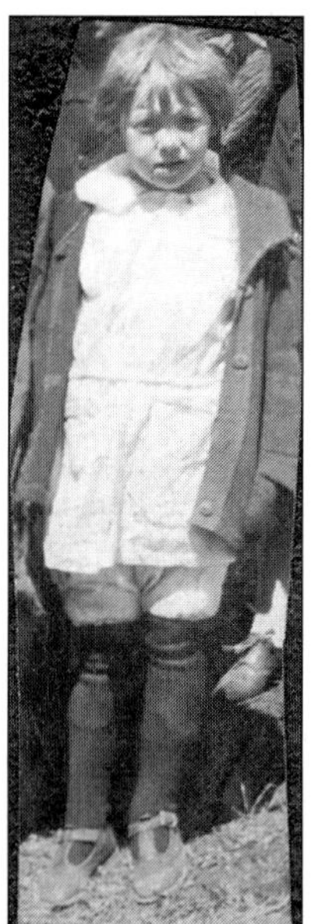

originally from Middlebury, Pennsylvania. He came to Canada in 1909 and found work as a locomotive engineer for the Fassett Lumber Company. His dream, however, was to become a farmer. He had recently moved back home to Pennsylvania with his wife, Grace Gaston, from Greece's-Point, Quebec, and their children. In Pennsylvania he bought a farm only to have his dream washed away by floods and storms. "Mother said 'that's enough of that, I'm not staying down here,' so he moved her back to Fassett," recalls their daughter Margaret. The family was on the move again, but now they were going to the new location of Allan's employer in the backwoods of Ontario.

When the train left the north boundary of Algonquin Park and passed by Kilrush Lake, it reached the top of one of the steepest grades on the line. It crossed an embankment and pulled to a stop at a tiny clapboard shoebox station perched on the slope overlooking the Wasi River valley at Fossmill. When Allan, John, and Teddy stepped down from the coach and had a first look at their new home they saw the emerging framework of new buildings.

Building the mill

By the spring of 1924 Fossmill was a hive of activity. When the transfer to Ontario was confirmed, Jack McGibbon brought his wife and three children to live in a rented log house on Wasing Road near the future mill. He worked with Tom Howard, from Huntsville, who was hired as the Local Manager. Roy Summers came from Quebec to oversee the finances and a local man, Alton

Hot pond: *Workers enlarge Foster's log pond.* [DS]

Douglas, ran the company office. Sidney Staniforth visited regularly.

The new mill was soon designed and construction was started. Carpenters and other workers came from Quebec and, along with some local men, built the mill, tramways, and outbuildings.

"Foster's old mill was little and most of the buildings were torn down. Most of them were pretty shabby," recalls Jack's daughter, Peggy. She was on-site from the early days and recalls that "The mill was so small my father didn't call it a mill." Fassett built its new mill on the site of Foster's mill. Workers enlarged Foster's log pond and put a steam engine and boilers into position. By the time John and Allan Gleason arrived workers were framing the new mill. John and Allan also saw carpenters

Mill framing: *The mill was built from heavy timbers. The framing in the centre of the roof is for the sawfiler's loft.* [DS]

Boilers: *The steam plant of the mill would be housed beside the mill in a metal covered building.* [DS]

Jack McGibbon: *This early photo shows the original Fossmill CNR Station in the rear and the boarding house on the right. Irene Howard, daughter of Local Manager Tom Howard, is in the middle, and Jack's daughter, Peggy, is seated in the car. "Even as a little kid I was watching and looking at how everything went on," Peggy recalled in a 1996 interview.* [CWI]

building a huge stable just west of the pond on a flat spot by the river.

Company carpenters built large well-constructed homes for Tom Howard and Jack McGibbon across the railway track from the Fossmill Station on high ground overlooking the town and mill complex. A new company boarding house big enough to house 50 to 60 workers in single rooms was built behind the station. It also had a large dining room for the workers and an apartment for the cook and his family.

Other carpenters built the company store and office next to the boarding house. Private rooms were built above the store for company officials and office staff. The telephone company, with the assistance of Fassett, installed a telephone line to Fossmill. This completed a telephone system that came from Pembroke along the CNR.

New construction: *The sorting shed and tramway. A snow plow sits in the yard ready for use.* [DS]

Local Manager: *Tom Howard, from Huntsville, was hired to oversee the mill and village.* [DS]

Hotel, office, and company store: *The office and store were on the first floor and the private rooms were on the second.* [DS]

Just finished: *Newly arriving workers from Quebec were accommodated in the new boarding house.* [DS]

CNR Switch: *Fassett locomotives at the switch from the main line CNR. [DS]*

Railbed: *Newly prepared railbed awaits ties and rails at the bend around the future lumber yard. [DS]*

New home: *This typical Fossmill house was for the family of Morris Donahue, the Fassett railway superintendent. [LN]*

The Fossmill logging railway

As locomotive engineer, Allan Gleason would have noticed the siding and switch just west of the station where the new Fossmill logging railway connected to the main line of the CNR. From there its rails descended gently to the south-east into the valley toward the mill. Beyond the mill the line curved, following the Wasi river, and disappeared between the hills south of Fossmill. Allan noted the steep grade and was glad his locomotive was a Shay, built specially for those conditions. John Gleason recalls early rides on the railway. "I was supposed to stay with this Mrs. Short who ran the boarding house. When I saw her I thought that was enough, I wasn't staying there. The only thing Dad could do was take me and Teddy with him. He put a seat behind the old engine and I sat there with the dog underneath the seat and we looked out the window as we went along, coming over the mountains. There were a lot of mountains."

New Houses

A flat spot just west of where the company railway line crossed the road from Wasing into Fossmill provided an ideal place for the Gleasons to build their new home. "We stayed in the boarding house while I helped my Dad build that house," John Gleason recalled in a 1996 interview. "Dad took the night job looking after the engines. I guess it was the only way he could get a house quickly."

By the time John's older brother, Lester, and sisters Nora and Margaret arrived with their mother three or four months later, Allan had built a substantial one-and-a-half-storey house with two bedrooms downstairs and two more upstairs. A verandah wrapped around three sides of the house, and there was a large front and side yard for a garden. There was no running water or electricity, and the toilet was an outhouse in back.

Other workers also came early from Fassett to Fossmill to work and build homes. Steam loader operator Jack Huneault came with his father Tom. Jack's brother, Donat, recalled the day the rest of the family arrived in Fossmill. "My Dad had the house pretty near built; all they had to do was put in the doors."

The MacDonalds

One of the local workers hired by the company was Gordon MacDonald. Born in 1890, he was one of the 10 children of Edmund and Charlotte MacDonald who came to Chisholm Township in 1886. In 1910 Gordon married Myrtle Welch, a native woman from Monteagle Valley near Bancroft, Ontario. Before working for Fassett, Gordon worked in the local Baechler Lumber Company logging camps, and Myrtle worked there as a camp cook.

In 1924, Gordon and Myrtle moved to a big house near the Gleason's and Gordon started work as a brakeman on the logging train. Gordon's daughter, Elva, worked as housekeeper for the McGibbons. Gordon, Myrtle, and some of their 13 children and grandchildren, became central characters in the Fossmill/Kiosk story.

Wasing school

In September 1924 the children of Fossmill prepared to go to school. Since there was no school in Fossmill itself, they had to walk the company road or the CNR tracks the two and a half miles (4 km) to the Wasing school (School Section 'SS' No. 4). Peggy McGibbon

Myrtle MacDonald

Gordon MacDonald

MacDonald children in 1924: *Back row left to right: Boyd, Milton, Irving, and Elva. Front row left to right: Freeman(Chick), Jean, Vera, and Irene.* [MC]

Brothers: *Amedee and Oscar Laferriere from Chisholm were hired to work at the new Fassett mill.* [GL]

remembers being taken to school in a cart that hauled scrap wood at the mill. "Those dump carts were great. We went to school the first year at Wasing in one pulled by a horse. The carts were big and sturdy. We thought it was a lark."

Peggy also remembers the shock the rural community of Chisholm felt because of the influx of students. "Wasing was just having a fit. Can you imagine what families at Wasing felt like with all of these kids coming?" Wasing student Doreen Smith (Litkee) recalls that "All the Fossmill kids came down to our school—there was three to a seat!" Nora Gleason feels that there was some bias, at first, against the new people from Quebec "until they got to know us—we thought they were a little odd too," she says.

An added complication was that many of the students from Quebec spoke only French. In 1912, Ontario enacted Regulation 17, which banned French in Ontario schools. "It was pretty hard for us," recalls Donat Huneault. "We went to a French school in Quebec. When I moved to Fossmill it was a different story; it was all in English." He remembers the help he received from a new friend. "It was all different for a while but I had a chum, Boyd MacDonald, he taught me a lot of English, that's how I learned." Germaine Demers (Danis) remembers the experience, "It was hard, in a way, because we didn't know English at all. I was laughed at lots of times."

In spite of Regulation 17, the English speaking students also got a taste of French from their new-found friends. "When I came to Fossmill I was about nine or ten," recalls Louis Demers. "I couldn't speak English. The

Wasing school, SS No. 4: *Near Fossmill.* [CWI]

Cold winters: *Fossmill students had to dress warmly for the long walk to the Wasing school.* [NG]

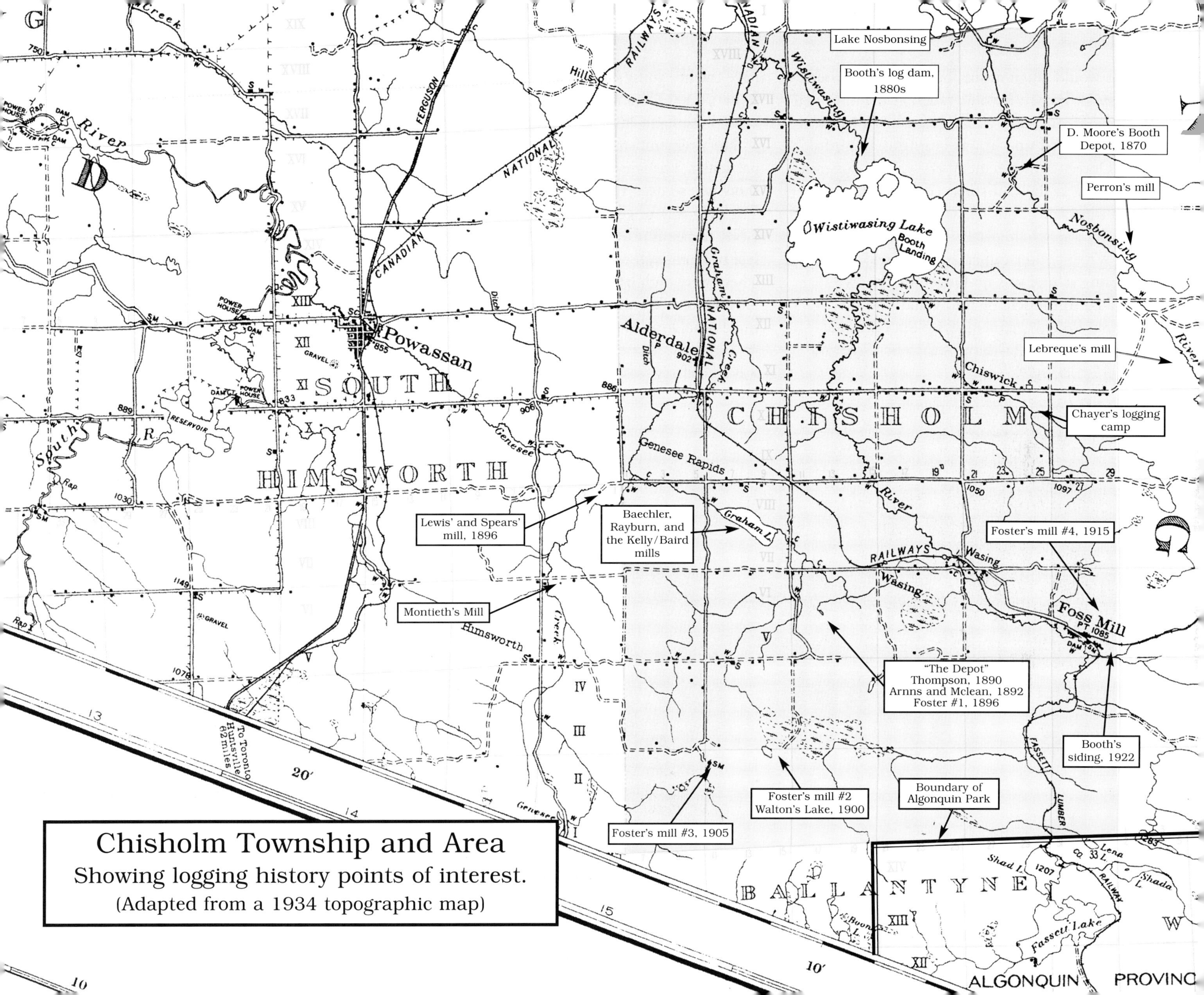

Chisholm Township and Area
Showing logging history points of interest.
(Adapted from a 1934 topographic map)

kids would laugh at me, but when they tried to speak French I would laugh at them. We all spoke French, and we all spoke English. We got all mixed up—it was beautiful, beautiful!"

Doreen Smith remembers when a Fossmill girl sang at the Christmas concert. "She had a beautiful voice, but she sang in French. It was lovely. The whole place was silent."

When winter came, the long trek to the Wasing school was no longer a lark. "We walked to the school and would find the school locked," remembers Nora Gleason. "The boy that was supposed to be there and have the fire going had chores to do first, so sometimes we had to wait until he came and started the fire. Oh, it was cold."

Nora remembers one winter morning in particular. The wind was howling and the snow was so deep the roads were impassable. The only way to get to school was on the CNR tracks. Railway plows had cleared a path just wide enough to let the train through. There were high banks of snow on each side. "It was cold. We walked together; there were three MacDonalds and two of us Gleasons."

Suddenly, near disaster struck. "All I knew was somebody pushed me and I landed in the snow drift. I thought it was Boyd MacDonald. I was going to give it to him because I landed hard. Just then a train came whirling past—then I realized what he had done."

At last: *Fossmill school, SS No. 4B.* [TP]

The Fossmill school

It was soon clear that the Wasing school could not handle the large number of Fossmill students. In 1925, the company petitioned the government for a school in Fossmill. Wasing school became SS No.4A and the company built SS No.4B in Fossmill. One year of overcrowding and long treks to Wasing was enough. Donat Huneault says, "We were darn glad when the company built the school."

The company built the Fossmill school on the side of the hill just east of the Gleason's house, central to the community, and handy to most of the students. "We could run home and grab a cookie whenever we felt like it, we were that close to the school," remembers Margaret Gleason (Frederick). The school was a simple framed clapboard structure with windows on most of the south side and a small cloakroom by the front door. There were two outhouses off to one side of the playground.

"I used to be the one to light the stove in the school," says John Rouselle. "The stove was a big box-stove like a barrel and stood in the middle of the room with a big pipe going this way and that way. I went there at eight o'clock, and by nine you could open the window. The kids sitting next to the stove would roast, and those sitting further away froze."

The Fossmill school had 45 to 55 students

per year, considerably more than the Wasing school. The grades ranged from grade one to high school, and all the students were in one room. The stories of the methods used by teachers to maintain discipline, and to teach the Fossmill students—as well as the students' response to the teachers—are legendary.

The first teacher was Doris Bannister, a young woman who had recently graduated from the North Bay Normal School (Teachers College). She had a difficult time controlling the older students. "There were a lot of big boys. She couldn't handle them," recalls Margaret Gleason, who was only six years old at the time. "Some of these boys, they were pretty close to being out of school. We were such a close-knit bunch in there that when she tried to give me the strap or anything—I was only a little thing— those boys would step up and say 'you are not going to touch her' and the teacher would start to cry."

That group of boys included 10-year-old Donat Huneault and 11-year-old Boyd MacDonald. Donat recalls, "I was with Boyd all the time and I was always in some mess. We were a bad bunch. I think about all those things we did in school and I feel sorry. We shouldn't have done that. We didn't touch her

Fossmill gang: *A group of Fossmill students in 1926.* [NG]

First teacher: *Doris Bannister with two of her students.* [NG]

but she cried all the time. She used to go in the cloakroom and cry there. I'm sorry for what we did to that first teacher but it's too late now."

The following year the school board hired Ted Priest, the son of a Baptist minister, a violin player, and former boxer. But more about that later.

Religion

Religion played an important part in the life of Chisholm Township and Fossmill. The Protestants built Anglican, Methodist and Presbyterian churches, formed an active Orange Order, and built an Orange Lodge. Fossmill Protestants regularly attended the Wasing United Church a couple of miles from Fossmill, and mixed with the people of the area there.

By 1910, over 90 Catholic families lived in Chisholm when the St. Louis de France Church was built in the village of Chiswick, about five miles (8 km) north of Fossmill. Since most Fossmill residents were Catholic, Father Joseph Gravelle, the local priest, travelled once a week to Fossmill and held Mass in the boarding house.

Fossmill's Catholic parishioners—apparently with the help of Protestant Orangeman Gordon MacDonald—eventually built a satellite church in Fossmill, east of the school. "The church was very plain, just regular," remembers one villager. "It had a big bell on it and square windows with little glass panes. It had double doors and seven or eight wide steps that went all the way across the front. It was in a low spot below the ridge where the company road went."

"The church was built on cedar posts six or seven feet high and the lower area was boarded round. A great big furnace was under there," recalls Roger Bergeron. "My job was to light the furnace the night before church, in the winter. I used trimmings and slabs from the mill. The furnace would take four-foot long pieces. You had to be careful, it was all stove pipes and sheeting up to the floor grate, which was about four-foot square. If you lit it in the evening, by the next morning, when it was time for the service, the church was nice and cozy."

Fossmill Catholic Church: *Scaffolding surrounds the Church as it is being built. Carpenters have just begun the top of the bell tower. [LN]*

John Rouselle had another memory. "I used to be an altar boy and one Sunday the priest was late. We heard a noise in the back of the altar so we lit a candle and looked in. There was a starlings' nest in there, full of hay. In our excitement we dropped the candle and set the gol-darn church on fire. When the priest came in everybody had pails of water. We put the fire out but it made a big hole. The priest didn't say anything but I got a kick in the arse at home for that. That's the way they raised us then."

The village of Fossmill was quickly becoming a real community. Soon more families would move in, and more houses and mill and railway buildings would be built, but the life blood of the operation was the supply of logs that would come from the limits in Algonquin Park.

Getting down to business

The December 1924 issue of the *Canada Lumberman* reported on the new plant in Fossmill. "The company hope to have their plant in operation early in January. They have...completed a switch [railway], connecting their plant with the woods operations and this line, which is five miles long, will be ample for the time being for the transportation of logs. It is the intention of the Fassett concern to operate its plant the year round, and their cut during the coming year will be between 8,000,000 and 10,000,000 feet."

In the next three chapters we will explore the operations and working conditions of the workers in the woods, on the railway, and in the mill. We will return to the story of the village and its people in chapter 6.

Steams up: *The hot pond is full of logs waiting to be turned into lumber in the newly completed mill.* [DS]

Early Fossmill: *View of Fossmill taken from the managers' side of the tracks, facing south. The first CNR station, boarding house, and company store are on the left. The old Foster buildings beside the tracks will be replaced soon with new homes, and an engine house, east of the mill, has yet to be built.* [CWI]

Logging: *Fassett lumberjacks show their expertise to students from the University of Toronto Forestry Department. [UTA]*

3
THE LOGGING SEASON

"Once the snow got deep, then they didn't come out as often."

Jack McGibbon cursed the cloud of blackflies swarming around his head as he looked at his compass. He counted his steps to keep track of the distance he had travelled and studied the forest around him. The rolling country around Moose Lake (now Fassett Lake) was covered with mixed hardwood and softwood. Jack also noted alder and cedar along the edge of a stream. Spruce and balsam grew on the flat areas at the bottom of the hills. He noted that a **hauling road** could be built there. Up on the side of the hill were pine and hemlock suitable for lumber. The bark of the hemlock could be sold to the leather industry for tanning. The upper slopes were covered in good-sized stands of beautiful virgin maple and birch.

Jack noticed the remains of huge stumps from white pines that J. R. Booth had cut for **squared timber** in the 1880s, as well as the smaller pine stumps left from Booth's second cut of **saw logs** only ten years before—no matter, his company was mainly interested in the hardwoods.

Cruising

Peggy McGibbon remembers that the first stage in the woods operation began every year when her Dad "started walking the limits in May and through June and July at the

Camp One: *The first Fassett camp was on Little Moose Lake just inside the boundary of Algonquin Provincial Park.* [UTA]

height of the blackfly season. Walking the limits is surveying, getting the lay of the land, and finding where you are going to cut and where you are going to put the roads to get the logs to where you want them. My father had an innate skill when it came to the height of land, the low of land, and the way water flows. He had the whole area topographically in his mind."

Woods Superintendent Jack McGibbon

A 1928 *University of Toronto Forestry Student Report* [5] explained how this walking or cruising was accomplished by the Fassett Lumber Corporation. "Cruising and estimating is done by the old-fashioned method. The woods foreman goes out, measures the distance by timing himself while walking, looks over the area to be logged, selects average acres, and, knowing the total area, estimates the total stand on the area from this. Mr. McGibbon, the man who does this work, is an old hand at the game and his estimates are very accurate, proving quite adequate for the requirements of subsequent logging."

The limits

The Fassett Lumber Corporation's limits included Foster's limits in the south-east corner of Chisholm Township and Booth's limits in the south end of Boulter Township, both below the CNR. As well, Fassett had acquired a section of Ballantyne Township in Algonquin Park from Foster. The richest part of the limits, however, were the limits in the Park that Fassett leased from the Booth Lumber Company. Starting at the west end of Lake Kioshkokowi, the limit's southern boundary followed the south bank of the Amable du Fond River, the west shore of Lake Manitou and the north shore of Tea Lake (now North Tea Lake) in Wilkes Township. Fassett also planned to acquire the rights to the still unopened limits south of Tea Lake in Wilkes Township.

Unlike the situation in Fassett, Quebec, where they had owned the land, the limits in Ontario were on Crown land, so the Fassett Lumber Corporation operated under the **diameter limit** system used on Crown land in Quebec and Ontario since 1882. This system stipulated that **loggers** could not cut down a tree that was less than 12 inches in diameter when measured close to the ground. In a speech to lumbermen in 1926, Sydney Staniforth said that his company "did not cut less than 12 inch trees." This system was thought to enable smaller trees to mature, ensuring a healthy forest in the future. Some foresters, however, argued that the diameter limit system, coupled with the taking of only **merchantable** timber, removed the best trees and left unhealthy trees standing, creating poor conditions for regeneration in the remaining forest.

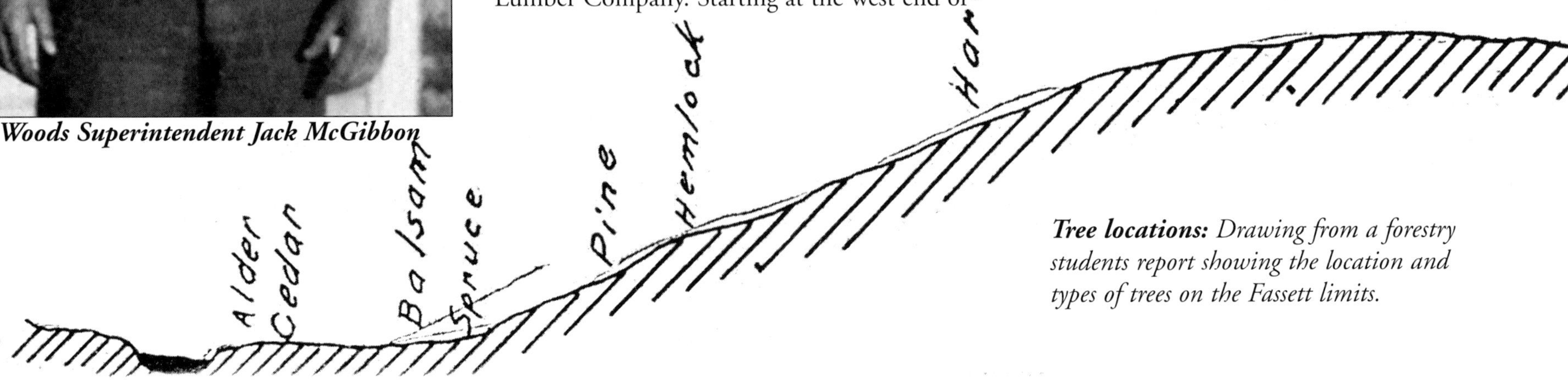

Tree locations: *Drawing from a forestry students report showing the location and types of trees on the Fassett limits.*

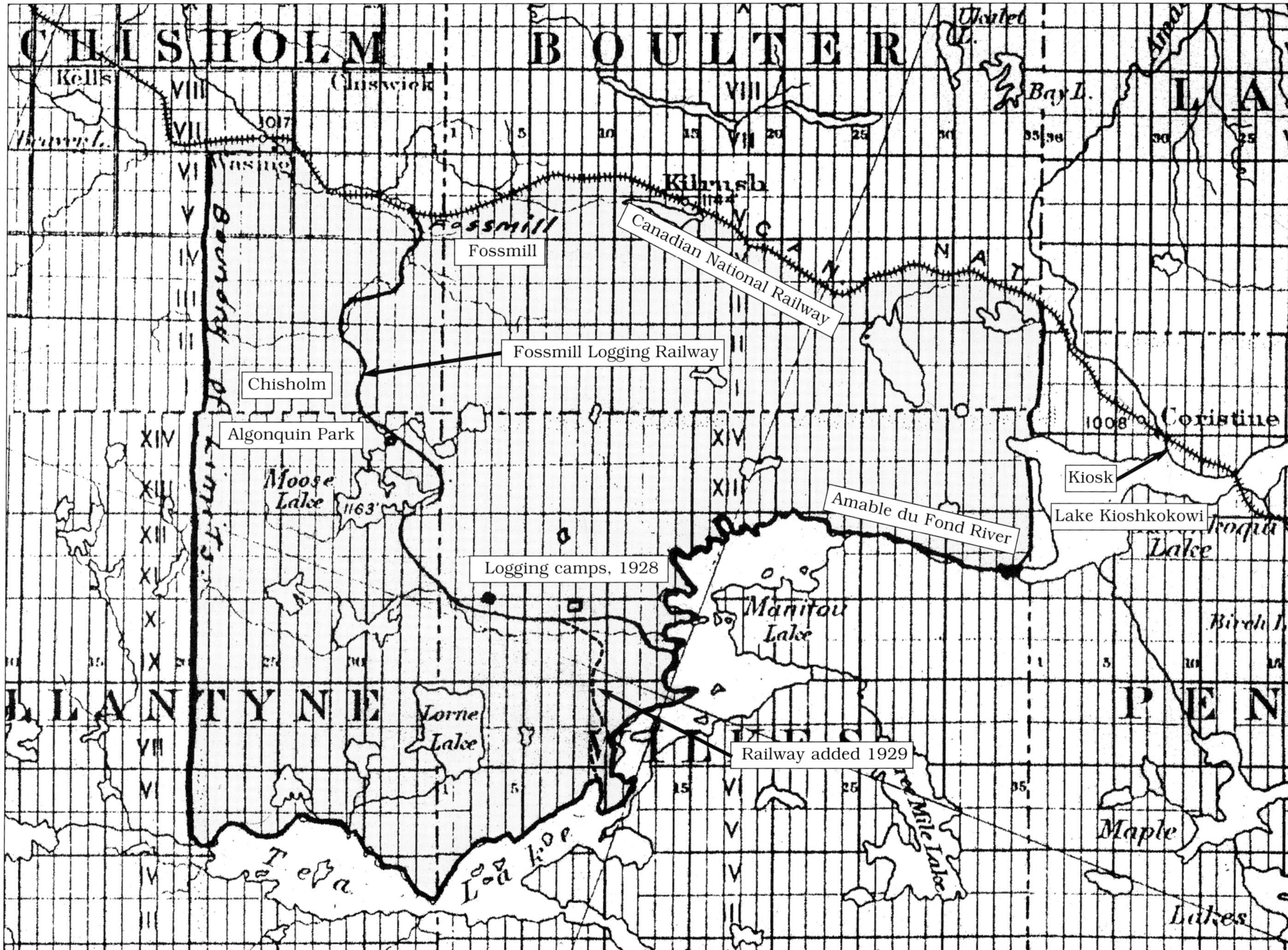

Fassett's Fossmill limits(shaded area): *This map is reproduced from a forestry student report, with enhancements and annotations.* [UTA]

Horse power: *Hundreds of horses were shipped in from "out west" on the CNR (above) to haul the logs in the bush (below).* [LN/UTA]

Men, horses, and snow

Another striking feature of the woods operations of logging companies in the 1920s and 30s was their reliance on "primitive" labour intensive methods. Their mills were "technologically advanced" but, as Ian Radforth writes in his book *Bushworkers and Bosses,* [6] "In their logging operations, even as late as 1945, lumber companies and pulp and paper firms clung to time-honoured methods that depended heavily on learning-by-doing, the strength of men and horses, and natural factors, such as the friction-reducing qualities of snow."

Fassett had invested heavily in its mill and railway. It required a huge number of logs to feed the mill to capacity and bring a profitable return on its investment. These logs could not be supplied by the manpower and horse power available in the local community. The mill workers worked year round and there were not nearly enough local farmers and teams of horses to staff the logging camps to capacity. To maximize production, Fassett hired jobbers from Quebec.

Jobbers

Jack McGibbon gave most of the contracts to the same jobbers he had used in Quebec because he knew that they were experienced, hardworking, and reliable. They brought their own crews for the season and went home at the end of the season when their work was done.

"The men that my father hired were from down around Thurso and Fassett," recalls Peggy McGibbon. "You remember the hockey player, Guy Lafleur? Two of his uncles worked for my father as jobbers. Jobbers worked independently. They were paid by the company but got their money because they did it well on their own. They cut the logs and hauled them to where they were told to put them."

The jobbers were paid for the volume of logs they cut. The contracts specified that they were to cut all merchantable trees, and leave standing the trees that were crooked, diseased, or too small. Once these trees were cut to standard log lengths they were hauled to **log dumps** beside the Fassett railway. The sleighs, horses, steam loader, and railway were owned by the company. The company store supplied the food for the men and horses, and billed the jobbers accordingly.

Not all the people who worked in the camps were French-Canadian. The 1928 student report, mentioned earlier, said, "There were a few Austrians, and a neighbouring camp, where they were cutting pulp and laying the railroad, was composed of Finns." Local contractors were also hired to help. Chisholm jobber, Joe Chayer, had a crew of about 30 men working at Moose Lake (now Fassett Lake) in the 1927-28 season. His lumberjacks included many well-known Chisholm men.

Horse power

Dozens of horses were also needed to haul and help pile the logs. Because there were not enough horses locally, the company imported them from Western Canada. Peggy McGibbon remembers listening to her father negotiate their purchase. "Harry Hamill was a horse dealer. He had been out West as a young man and had contacts out there. There were lots of men that did that. They were like brokers. They bought horses and then resold them. I sat and listened to them talk horses."

Lionel Martin, the son of one of the Fassett brakeman, remembers the scene at the railway when the horses arrived. "The horses would come in by the carload in the fall. They would unload those carloads and the horses were so lively—they were just kicking up their heels in the yard."

The newly arrived horses were taken to the paddock next to the stable beside the log pond. The stable was not large enough to hold all of the horses destined for the bush, so they were soon led along the railway tracks and

Jobbers' Contract

The main terms of the jobber's contract were:

1. Jobber agrees to take out
   a) All merchantable saw-logs
   b) All pulpwood (spruce and balsam) on a given territory.
2. Jobber agrees to cut all hemlock, spruce, balsam, cedar, pine, basswood and ash logs, eight inches and up at the small end, in 10, 12, 14, and 16 feet with equal proportions of 12, 14, and 16 feet logs; To cut all birch, maple and elm, ten inches and up at the small end, in 10, 14 and 16 feet lengths with not more than 65% 14 and 16 feet lengths; To allow an overrun of 2 to 4 inches on each log; To cut maple and birch 10 and 11 inches at the small end, which is sound and straight, 8 or 16 feet for tie timber.
3. Jobber agrees to place hardwood and softwood in different skidways along the main roads with the ends and sides of the skidways clear of brush.
4. Jobber agrees to deliver the same in dumps along the railroad within 75 feet of the railroad, and the softwood and hardwood piled in separate skidways. The dumps are to be cleared of brush and stumps; to deliver pulpwood to the railroad and to load on cars.
5. Doyle Rule is to be the log rule used, to accept the government scale as final.
6. Jobber agrees to buy all equipment from the corporation.
7. Jobber agrees to build all camps at his own expense and to use no good timber according to government regulations.
8. Jobber agrees to turn in, to the corporation, the payroll, on the tenth of every month.
9. Jobber agrees to accept medical attention and deduct $1 from each man's wages.
10. Jobber agrees to follow all the rules of the Department of Lands.
11. Jobber agrees to abide by all requirements of compensation board, and pay assessments.
12. Corporation agrees to pay an allowance on logs at the skidway and the remainder at the landing.
13. Corporation agrees to have logs scaled each month, and to pay the jobbers, on the 15th of each month.
14. Jobber agrees to do the work according to the rules of the corporation as interpreted by the log inspector.
15. Jobber agrees to complete all cutting, skidding and road-making by January 1st, and to deliver all logs and pulp by the railway by March Ist.

Reproduced From the Forestry Student Report on Fossmill.

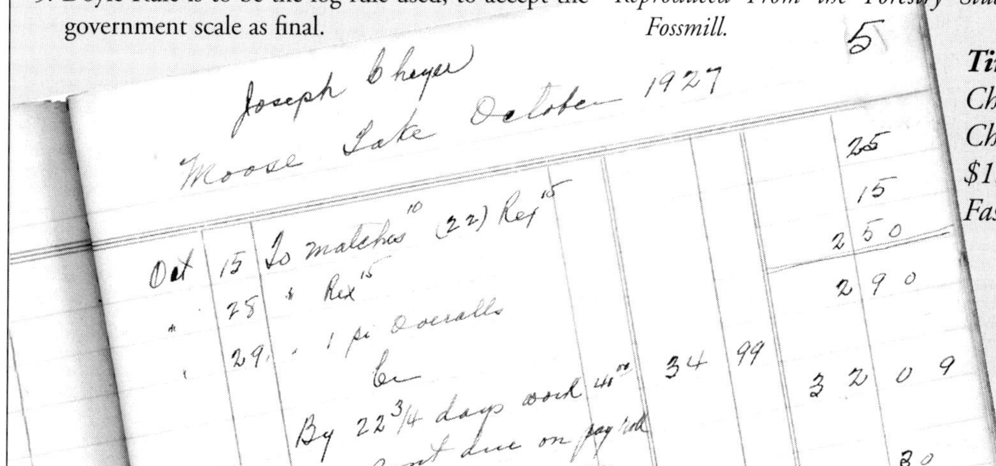

Time sheet: *Chisholm's Joe Chayer worked for $1.45 per day for Fassett in 1927,*

Supplying the camps: *One of the jobs of the Fossmill logging railway was to deliver supplies to the camps. [UTA]*

Winter home: *Contractor Leon Lafleur's camp. [UTA]*

roads to the logging camps.

Jack McGibbon did much of his travelling in the bush on horseback. His daughter remembers when he bought his favourite horse. "One day Harry Hamill told him, 'I have a beautiful mare for you.' Dad liked her, and bought her. The horse's name was Minnie—Mr. Hamill's wife's name was Minnie; I was young; I thought that was funny." Her father bought a "beautiful western saddle" for the horse, which wouldn't let anyone but her father ride it and wouldn't stop moving—"just like a race horse," she says.

Peggy recalls her father leaving to go into the bush. "He went down to the stable and got his horse and then came up to get whatever he needed in the store. The horse wouldn't stay still. We kids would run upstairs and look out the east window to watch him gallop away. Gosh, it was a beautiful horse!"

The season begins

In September, when the leaves on the trees were beginning to turn yellow and red and the weather was turning cold, the jobbers would arrive in Fossmill and head into the bush to establish cutting areas and campsites.

Once the road and campsite locations were determined, the jobbers brought in select crews to build the **tote roads** or **cadge roads** that were used to transport supplies and equipment into the campsites. A cadge road was a low quality utility access road and was not used for hauling logs. Log-hauling roads had specific requirements for handling the movement of loaded sleighs over ice and snow.

The logging train took supplies to the camps by the railway. Other camps were supplied directly from Fossmill. Chisholm historian Stanley Anderson, whose family farm was a few miles away in Chisholm, worked in the company store in the winter of 1928-29. Stanley recalls supplying the Desabrais camp, directly east of the mill. "There was a team that went up daily with supplies for their stores and their cooking. They called it a 'cadge team'. There was a road through the wilderness, a trail that was called a 'cadge road'. I assembled the stuff at the store and went down to the barn and got a few bales of hay and some oats for the horses. These cadge roads went up and down over the country by the shortest route. When the snow came we used the ice-roads that were used to haul the logs, rather than follow the top of the hills."

The camps

In 1928-29, three of the six Fassett camps were on the railway. The 1928 forestry student report describes in detail the camp of jobber Leon Lafleur from Lachute, Quebec. His camp was located near Beaver Creek and the railway.

The plan reproduced on the right is from the student report and shows this camp. The buildings were made of unmerchantable logs and the floors, gables, and roof were built from **cull** lumber brought from the mill. The roof was covered with tar paper held in place by **lath** strips. The camp accommodated 75 people, 44 in the large bunkhouse, 20 in the smaller one, and the rest in other locations such as the back of the office and the cookhouse. The forestry students reported that the large bunkhouse was for "French-Canadians" and the small one was for "Finns, Pollocks [sic], and other foreigners." The private "house" was occupied by the cook and his wife, who also did the laundry for those willing to pay her $1 per month.

The stable held 12 teams of horses and all of the equipment necessary to run them. The blacksmith shop made much of the equipment for the camp and maintained the sleighs with the help of the camp carpenter.

Blacksmith shop *[UTA]*

Leon Lafleur's Camp *[UTA]*

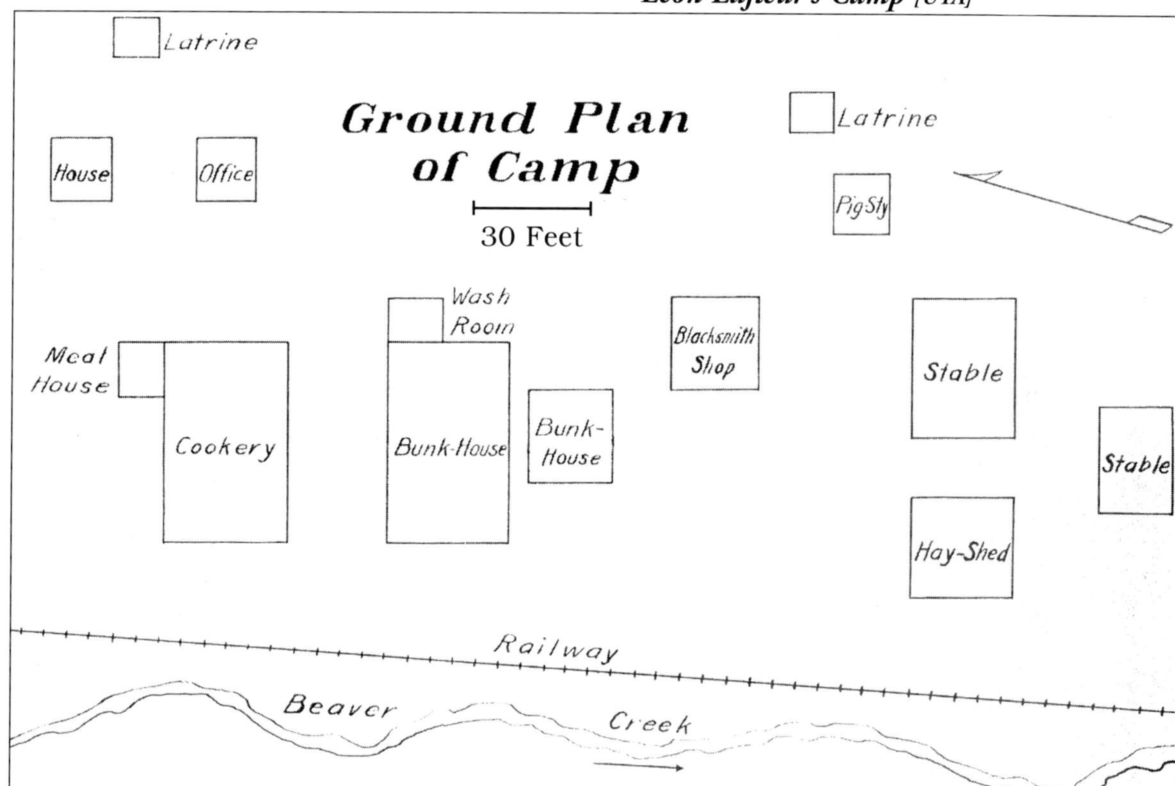

Mealtime

Going to work

Undercutting

Cutting down trees

When all the men were assembled and preparations were complete, it was time to begin cutting.

The lumberjacks began their day before sunrise with a hearty breakfast in the cookhouse. After breakfast, in the early morning glow, the men headed out to the cutting areas in the bush so they could begin to cut as soon as it was light enough. The foreman assigned teams of men to specific areas. Each team consisted of two **sawyers**, a **teamster**, a man called the **roller** or **decker**, and a trail-cutter. The decker worked with the teamster **skidding** the logs, to the log piles, which were called **skidways**. He also cleared brush and trimmed fallen trees. The trail-cutter cut 150 to 1000 foot (50 to 300 m) long trails called **snaking trails**, which zigzagged up and down the hillsides to get the logs out safely, in sometimes difficult locations. The student report noted. "Every effort was made to have snaking trails descending to the skidways, in order to save the horses."

The head sawyer of each team selected a tree to cut. A notch, called the **undercut**, was made with an axe in order to control the direction the tree would fall. The sawyers then used a two-man **crosscut saw** to fell the tree. The head sawyer decided the best way to cut the tree into logs, and, avoiding crooks and bends, marked the lengths with a measuring stick.The sawyers then **bucked** the tree into log lengths with a crosscut saw.

The head sawyer then selected another tree, and the process was repeated throughout the days and weeks ahead. Each team cut between 25 and 30 trees per day. Each camp cut an average of 16,000 trees on approximately one square mile (2.6 sq. km) of forest per season.[7]

When the logs were ready and the **trail cutter** was finished, the teamster came in with his horse and skidded the logs to the skidway.

***Logging life**: Scenes from the Fassett movie. (see page 151) [photos with rounded corners are stills from* Logging by Rail in Algonquin park *the video of the movie.]*

Crosscut sawing

Bucking

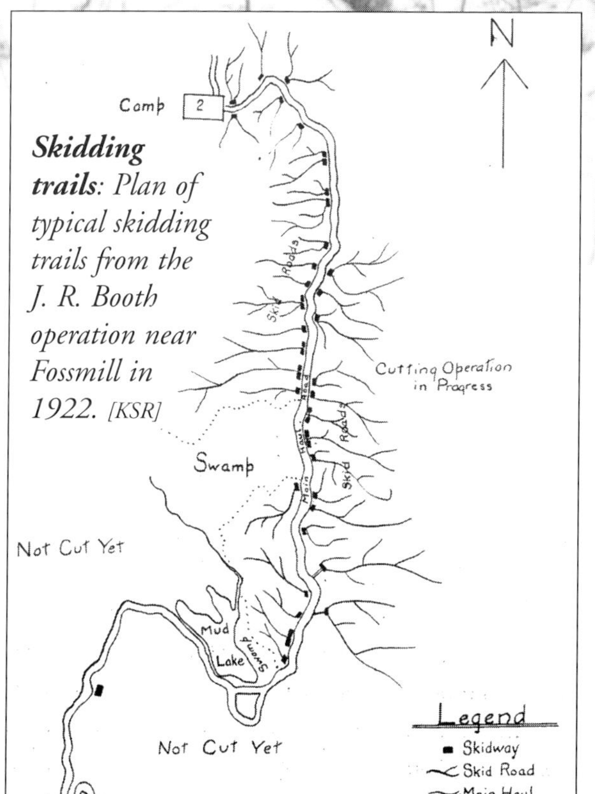

***Skidding trails**: Plan of typical skidding trails from the J. R. Booth operation near Fossmill in 1922.* [KSR]

The skidway was the temporary site close to the cutting area where logs were piled until the hauling roads froze, and the logs were taken by sleigh to the main log dump beside the railroad for later pickup by the Fossmill train.

The teamster and the skidway crew put the logs onto the log pile. This was called **decking**. They used a **decking chain** and two **skids**, which were long notched poles that were placed at right angles to the pile, so that the logs could be rolled up them onto the pile. The 1928 forestry student report described the process. "One end of the decking chain is passed underneath and then over the log to be decked and fastened to a log directly below where it is to be placed. The other end passes

***Decking**: Horses and a "decking line" are used to pile logs at the "skidway".* [UTA]

Skidding

Scalers at work

through a block on a tree at the rear of the skidway and back to the horse that pulls it. When the horse begins to pull, one man at each end of the log helps to roll the log and keep it straight until it is on the pile. They use a **peavey** to help with the job. The chain is then removed and the process is repeated with the next log that arrives."

The cutting and skidding was usually completed by late December. There was a break for Christmas while they waited for the hauling roads and lakes to freeze solid before beginning the main log haul to the railway.

Scaling

The jobbers were paid on the 15th of each month based on the volume of logs cut as determined by the company **scaler**. The scaler and an assistant used a **scaling stick**, to measure the diameter of each log at both ends, determined the length of the log and calculated by formula the number of board feet it would produce (see sidebar opposite). This was recorded on a **tally board** attached to the scaler's left arm. The scaler gave the information to the woods superintendent, who took it to the company office in Fossmill.

The company scaler went from camp to camp, sometimes travelling with the woods superintendent on his rounds. A good scaling team could record up to 2,500 logs a day. The team was well-paid, making $12 a day for the two men, with all expenses covered by the company.

The Fassett Lumber Corporation paid the Booth Lumber Company and the government a **stumpage** fee based on the volume of wood cut, and government scalers worked independently doing the same thing as the company scaler.

Wages

Wages varied depending on the job. The basic pay through most of this era was $40 per month, with specialists getting more. Top pay, not including the foreman, was $100 for the blacksmith, $90 for the cook, $75 for the sawfiler, and $45 for teamsters. Some sawyers and skidders were paid per 1,000 board feet log measure, as an incentive, and some did remarkably well under this system.

During the month, workers bought whatever they needed, other than food, at the camp store, or **van**. The cost of these purchases was subtracted from their monthly pay. Some men took advances to send home and some gave a monthly stipend to the church. One dollar was taken off of each man's pay for medical needs, and medicine was "free". The company paid a fee equal to 4.5% of each worker's pay for Workers' Compensation Premiums.

Workers who quit early, regardless of the amount of time worked, received a flat $10 (if they had earned at least that much). Workers were responsible for keeping the axes and crosscut saws they used sharpened and in good condition; they received a refund if they returned them in good condition.

"Doyle Scale Rule"

The job of the scaler was to approximate the amount of lumber that the mill would produce from a round log. This was determined by applying a formula to the measurements of each log in the skidway. One of the the formulas used was called the "Doyle Rule" even though it consistently produced a figure under the actual production in the mill.

The formula was:

> **Deduct four inches from the diameter of the log inside the bark on the small end, square one forth of the remainder, and multiply by the length of the log in feet.**

The scaler also had to take into account the condition of the log.

Broken ankle

Lumberjacks worked with sharp tools in dangerous and unpredictable conditions and were often injured far away from medical help. Max Smith recalls a story about Albert Atkinson, a local man who broke his ankle while working in the Park.

Albert's step-brother "brought him out 14 miles on a sleigh to catch the train. It was the middle of winter so they covered him with blankets, held down with bricks." The driver drove so fast that he covered the distance in an hour and a half. The horses had been driven so hard that they were covered in white foam sweat. "It was a black team and when it got there it was white," Max says. When they arrived in Fossmill, the 12 o'clock train was at the station, and they ran and put him on it. Albert's ankle was set in North Bay and he recovered and worked for many years as a piler in the lumberyard.

Home for Christmas

Workers occasionally went home briefly, especially at Christmas. Chisholm resident and local trapper Lloyd Tennant recalls that his father "used to go into the bush in October and work to the end of April, but he would come out some weekends. When they were in at Manitou Lake they came home across where my trap line is now. It was a four hour walk. They were working about two and a half miles west of the camp. The trail-cutter would take their team back to the camp for them and they would leave at 3:30 or 4 o'clock and head for home. Once the snow got deep then they didn't come out as often."

Camp conditions

The lumberjack's life involved long hours and rough conditions. Everyone was up each day in darkness, and came back to camp in darkness, only seeing the camp in daylight on

Cook out: *Lumberjacks take a break from their work for a hot meal in the bush.* [APA#3226]

Cookery: *The kitchen and dining hall at Lafleur's camp prior to dinner.* [CL/UTL]

Waiting: *Hauling sleighs stacked waiting for the hauling season to begin. A water tank is in the rear.* [UTA]

Sundays. Lunch was taken in a bucket if work was at any distance, and often consisted of cold salt pork and hot tea, consumed around a fire. If several workers were in one area, a hot meal would be delivered.

Camp cooks fed 75 to 100 men. They and their helpers, who were called **cookees**, spent many hours preparing food. In order to get the men out of the dining hall quickly, the cooks frowned on the men socializing during meals. In some camps, talking was prohibited while they ate. Food consisted of fresh meat, salt pork, eggs, sausage, beans, vegetables, lots of homemade bread, sugar, maple syrup, desserts, and all the tea and coffee the men could drink.

Good food was essential for a successful camp. The men worked in cold weather, expending a lot of energy. A 1931 report [8] calculated that a lumberjack consumed an average of five and a half pounds of food or 7,250 calories per day, over three times what an average man consumes.

During evenings and Sundays in the bunkhouse many tall tales were told and card games played. If there was a musician in the group, he was encouraged to the fullest.

Each man was given two blankets for his bunk. Uninvited guests to the camps in the form of bed bugs and lice were common and relatively uncontrollable. The forestry students made special reference to a sanitary solution practiced by a camp made up of Finnish workers. They built a steam bath or sauna in which stones were heated all day. The report said, "At night, the men assemble in the 'Bath House' and pour water on the stones filling the place with steam. The men lie on the shelves until well 'steamed', and then rub themselves down, vigorously with towels."

During times of high employment some workers would not tolerate poor conditions and moved on quickly. The student report indicated problems with **jumpers**, workers who quit work and moved to other camps. "Many men who were out of work would take work and work until they were well fed and then, on account of the nearness of the railroad and the ease with which they could get out, would jump." Labour historian, Ian Radforth, however, suggests that jumping was a form of protest against poor working and living conditions in the years before unions. [9]

One Chisholm farmer remembers going into the bush with a friend and a team of horses. He remembers the hard work and long hours. After three weeks in the camp his friend asked what the itching was, lifted his shirt, and found he was covered with bed bugs. They collected their $10 flat separation pay and left the next day.

It is estimated that during this period, 20,000 to 30,000 men worked each winter in Ontario logging camps. For most, the long winters away from home and family were balanced by the satisfaction that the bills would at least be paid that year, and at some time in the future they would be able to stay at home on the farm and work there.

A *Canada Lumberman* article complained that many good men "prefer remaining in the bosom of their families, providing they can get steady work, to undergoing voluntary exile in the shanties." Anyone, however, who spends any time in the bush in Ontario, especially in the fall, will understand the experiences of the men in the bush were not all negative. Ian Radforth quotes some commentators on the logger's life. One of them wrote, "Camp life was like a fever that some could not escape." Another wrote that there were men who "not being able to get along with their mates, sought solace in the company of men, hard work, good food, and the enjoyment of outdoor life."

Woods Superintendent Jack McGibbon had the luxury of a private office and sleeping quarters in the bush and was paid around $150 per month, but the demands of supervising all the camps kept him away from home most of the time. His wife, Stella, remembered that he spent all his life in the bush only coming home on weekends, "and sometimes—not on weekends—he was away, all of the time," she said. Jack's only time away from the job was during a few weeks in the summer when the family usually went to his family's home in Lachute, Quebec.

The log haul

After the short break for Christmas, and when the hauling roads were frozen solid, the hauling of logs to the dumps began. Some cutting and skidding continued, but the emphasis was on getting the logs cleared from the skidways and delivered to the dumps at the railway.

The hauling roads were made as flat as possible by filling low spots with rocks and brush and covering them with logs and fill. Material from the high side of a road was

The hauling season begins

Teamster moves a sleigh into place

Sleigh carefully loaded with a decking line

Down hill: *Hot sand and careful driving bring a load down a grade. [RD]*

pushed to the low side to create a level surface. Bridges and culverts were seldom used. The road builders were called **swampers** or **beavers** and the foreman the **buck beaver**.

In 1928, Leon Lafleur used 60 men divided into four teams to haul the logs. The teamsters now worked with a pair of horses and a sleigh. A four-man team, composed of two **senders**, one **toploader**, and one **tailer**, loaded the sleighs

A decking chain was used to load the sleighs. The sleighs were just under 12 feet wide, but, by careful placement of the bottom outside logs, the sleighs were able carry more logs. John Gleason recalls how the loading team "would put the big logs on the sides. There'd be an extra foot sticking over the sides. That way they could get a load 14 or 15 feet wide."

The sleighs were specially designed to carry these heavy loads of logs over the uneven hauling roads. They were in two sections called **bunks**. The front runners were joined to the back runners with two **cross-chains**. As the name implies, each chain crossed to the runner on the opposite side of the sleigh. This allowed each bunk to adjust independently to the road, and to negotiate curves.

The roads were plowed, iced, and made ready for the dozens of trips ahead. During the night a **sprinkler tank**, mounted on a sleigh, was used to cover the roads in a fresh layer of ice. Working by lamp light, the men filled the tank from a hole in the ice on a lake or river. A horse pulled a barrel of water up and tipped it into the tank. The tank was pulled over the roads, and the water was sprinkled from the back so that it would freeze behind the tank. The additional ice smoothed and increased the stability of the roads. The hauling roads were kept clear of debris by workers called **chickadees**.

A downhill run was a challenge to horses and teamster in icy conditions, and there were regular close calls. Workers called **sandpipers** heated stockpiled sand and spread it on downgrades to slow down the loads. One worker recalls that the sleighs' "runners were

Lakeside rendezvous

Sprinkler tank being filled with water

Truck hauls logs on lake

Frozen lakes: *Cletrac 30 tractor, log sleigh, and a sprinkler tank working on a lake on the Fassett limits.* [RD]

about six inches wide. They had an inch-thick shoe underneath made of half-round soft metal so they would catch on the sand when they came down the hills."

After the logs were unloaded at the dump, empty sleighs returned to skidways on another road appropriately called the **go-back road**.

Wherever possible, logs were hauled across frozen lakes, which were flatter and smoother than hauling roads on land. Often a number of sleighs were linked together for the trip across the lakes, especially with the introduction of trucks and tractors, which could pull larger loads.

Tractors

The Fassett Lumber Corporation was progressive in the use of new technology. By 1927 logging tractors became available and the company purchased four Cletrac 30s, made by the Cleveland Manufacturing Company. They could make six trips a day across lakes, pulling several sleighs and replacing numerous horses.

Photographs of the Fassett tractors were featured in Cletrac advertisements at the time (see over). At a lumbermen's meeting in 1928 Sydney Staniforth defended the use of tractors. Commenting on the "up-keep of horses and the danger of their loss through accident and drowning," he said that his use of tractors had been "economical and satisfactory."

Tractors occasionally had problems with the ice on the lakes, in spite of the regular use of sprinkler tanks to thicken the ice. "Those **caterpillars** were a no-good outfit," declares Peggy McGibbon. "They worked, but they were too heavy. All the weight was located in one spot. They were always going through the ice. If you want some fun, try figuring out how to get one of them—fastened to loads of logs—out of the lake. That was my father's job, and he always said, 'give me horses any day.'"

Through the ice

"Of course, that's an awful weight," one worker recalls. "They drew six or seven sleighs behind them. You're talking about a lot of weight. I've even seen horses go through the ice. It was a lot of fun out there. We never

August 1, 1927 CANADA LUMBERMAN 189

Illustrated above, Cletrac Model "30A" hauling 6000 feet (Doyle) green birch logs on operation of Fassett Lumber Corporation at Fossmill, Ont.

Cletrac

CRAWLER TRACTOR

Many woods operators in considering their hauling problems are sometimes at a loss as to what plan to adopt as a means of transportation for their logs—Tractor or horses—

Frequently the operator—cutting only a million feet needs a tractor as badly as does the operator cutting twenty million feet—both have heavy grades, long roads and lakes.

The operator cutting one to five million feet a year does not want to make a heavy investment and consequently has frequently purchased an agricultural tractor for hauling his wood because the investment was not great—**failing to realize that for very little more he could obtain a Cletrac Crawler Tractor** containing more power, better materials, etc. and which can be relied upon at all times. For winter operations, in order to be reliable, a tractor must have—in addition to power, weight and quality of mterials—a circulating cooling system by means of a pump—a high tension magneto—a centrifugal air cleaner in place of a water air cleaner. Cletrac has all of these —and the Cletrac Model "K" 2-ton having a rated D.B. horsepower of 15, **and an actual drawbar pull of 24.53 H.P., and 4590 pounds. Costs only** $2050.00 f.o.b. Toronto or Quebec—Duty Paid.

MODEL "A" CLETRAC
(Illustrated Above)

30 H.P. Rating **Weight 7600 lbs.** **Delivers 38 Draw Bar H.P.**
***Price:* $3800.00 f.o.b. Toronto**

Catalogue mailed upon request.

J. S. INNES

MONTREAL
Mount Royal Hotel

TORONTO
1109 Federal Building

QUEBEC
71 St. Peter Street

SERVICES AND EQUIPMENT

Road Layout and Construction—Skidways and Forwarding Yards

PLOWING — TANKING — LOADING — HAULING

Logging and Side Road Tractors—Logging Sleighs—Tank Sleighs—Snow Plows
Steel Cable—Barienger Brakes—Log Loaders—Skidders
Barton Portable Pumps

knew when the ice was going to break. Jack McGibbon lent me his big buffalo coat one time. He said, 'You better put this on, you're driving across the lake. It's damn cold.' We were coming across and the damned ice broke right underneath us. The tractor just got over the break but the sleighs went down. The **brakee** couldn't press the lever that pulled the pin out to let the tractor go. He yelled for me to stop but I couldn't hear because I had that big coat on."

As the ice collapsed underneath the sleighs and the water rushed in, the tractor pulled the sleighs over the opening, dragging them through the water and ice. A second brakeman was on the back of the load and "when he saw the water he jumped off the back and he couldn't get going for falling and slipping. We pulled them all through, and I tell you we were all shaking. We split the loads off and got the heck off the lake. Two days later we were on the lake again. God, we were driving across the lake, right over where we went through—the ice froze right up again."

When the sleighs arrived at the dump by the railway, the steam loaders unloaded and stockpiled the logs into separate hardwood and softwood piles beside the track until they were needed at the mill.

Spring

By middle or late March the haul was complete, the snow and ice were melting, and the skidways were empty. The two University of Toronto forestry students reported on the condition of the cut-over forest in the spring of 1928. They noted that all "merchantable saw-logs are taken off. The unmerchantable material is left standing." All that was left of the maple were "seedlings to saplings up to about two inches in diameter." They suggested that the new growth was "of sufficient density to make a good stand if it matures." They predicted that the unmerchantable trees left standing would seed bare areas, but that "The seedlings from them will not be as vigorous as those from sound and healthy trees."

In order to eliminate the costly hauling by road, some softwoods were dumped into lakes in the spring, tied into **booms**, and floated across the lakes to the railway (see aerial photo page 78). The bark from hemlock trees contains tannic acid, and was called "tanbark" because it was sold to tanning companies to preserve leather. "Little Moose Lake (now Shad Lake) was polluted with acids from the hemlock trees," remembers one avid fisherman. "The fish up in there were swimming around and couldn't see you. It affected their eyes. You wouldn't eat those fish; you wouldn't even catch them. Years later, when the company moved on farther into the woods, the fish started to come back again."

With the coming of warmer weather in the spring, the melting snow turned the ground to mud and bog, preventing further work. The jobbers settled accounts, closed up camps, and sent the lumberjacks home to their families, or to summer employment elsewhere.

Doreen Smith remembers when her father and her brothers returned from their winter in the bush to their farm in Chisholm. "They came home very unwashed, usually with head lice and bed bugs. They didn't get in through the door of the house. My mother would fill a big old wash tub in the woodshed with hot water. They took everything off and bathed. The clothes were left out, so none of that stuff would come in."

Another familiar spring sight was the spectacle of the horses being brought out of the bush. "They came by the dozens," remembers Peggy McGibbon. Teamsters moved the horses in stages because the stable at the mill was too small. Each group of horses was quickly moved to summer pasture. Peggy remembers that they walked "the horses from Fossmill to either of two ranches, which were big open fields. There were two men hired to live at these ranches and to look after the horses until the next fall." Local farmer Max Smith recalls that "They used to take the horses to Boulter Ranch every year. They'd drag them down that road. We were just young kids and we could see them going with these hundred horses all in a row." Lionel Martin remembers the "lively" horses that were "kicking up their heels" in the fall and how the "same horses came out in the spring, and there was almost nothing left in them; they were just dragging themselves."

Soon, Jack McGibbon would be heading back into the bush to curse blackflies and plan the following year's cut. Thousands of logs were now piled beside the railway in the Park—waiting for the Fossmill train to take them to the mill.

Summer vacation: *Company horses were taken to summer pasture at the "ranches" in Boulter Township east of Fossmill.* [CWI]

Shay-geared locomotive No 54: *The workhorse of the Fossmill operation.* *[LN]*

4

THE FOSSMILL RAILWAY

"... a rough, rickety railroad as railroads go."

Blasting white steam and black smoke, locomotive No. 54 approached the top of "Split Rock Hill". Engineer Allan Gleason prepared for the steep descent to Fossmill. His son Lester, the fireman, checked the water level and topped up the boiler so that the crown of the fire box would not overheat when the locomotive angled down the hill.

Allan blew the steam whistle signalling the brakemen, Emanuel "Man" Martin and Gordon MacDonald, to leave the caboose and start the treacherous walk across the top of the loaded log cars. Jumping from car to car they carefully turned the wheels that applied the brakes on each car. This dangerous procedure slowed the cars, so that their momentum would not overtake the locomotive and force it off the rails. Man and Gordon wore **calk boots**, boots with short spikes through their soles, to grip the pitching and swaying logs. Just to the side of the railway, at the top of the hill, was the huge boulder, split by the ice, that gave Split Rock Hill its name.

Five miles (8 km) of boulders, swamps, valleys, steep hills, and forest lay between Fossmill and the logs in Algonquin

***Log cars:** A Barnhart loader and log cars parked by the lumberyard are a source of interest to some Fossmill visitors.* [DS]

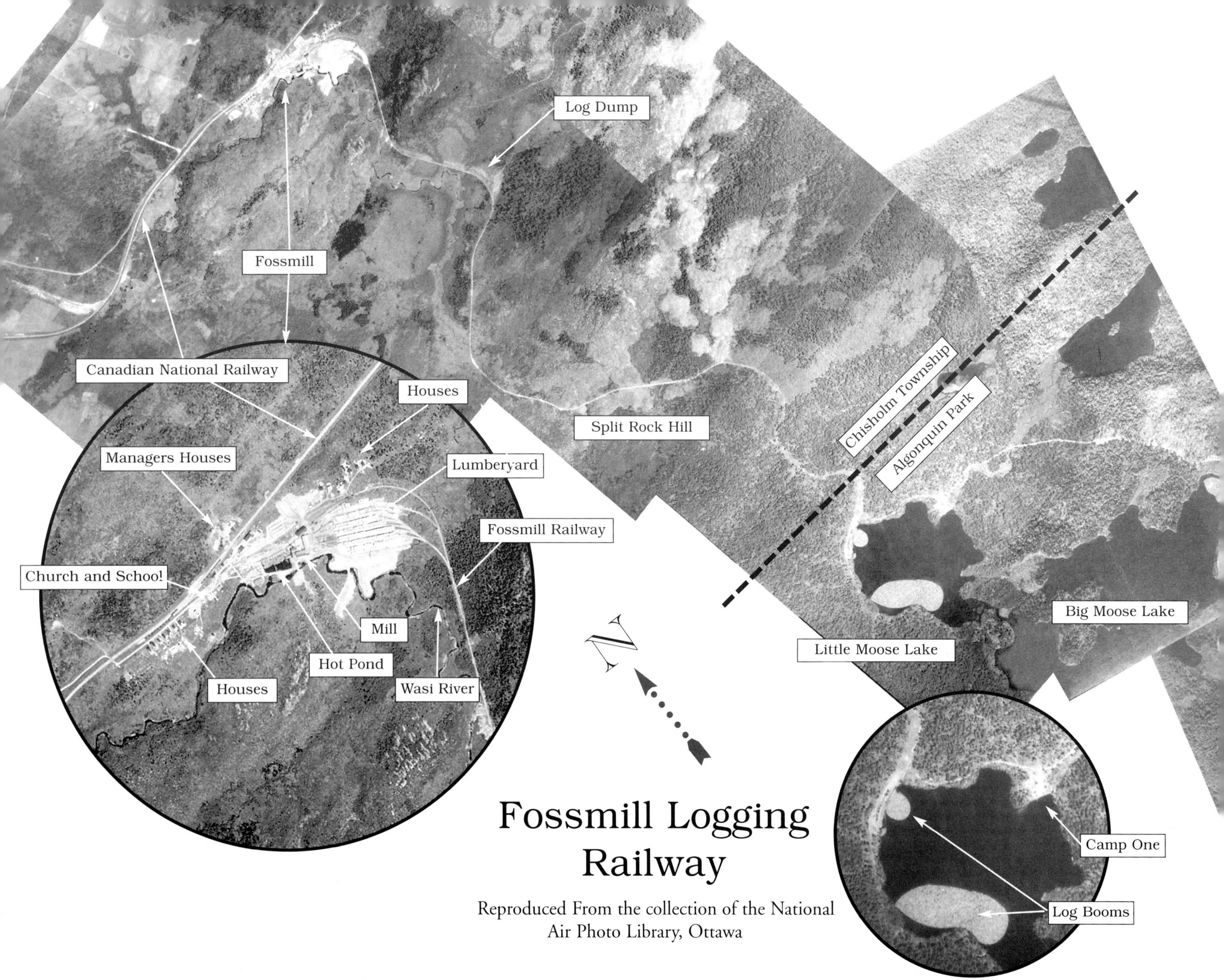

Fossmill Logging Railway

Reproduced From the collection of the National Air Photo Library, Ottawa

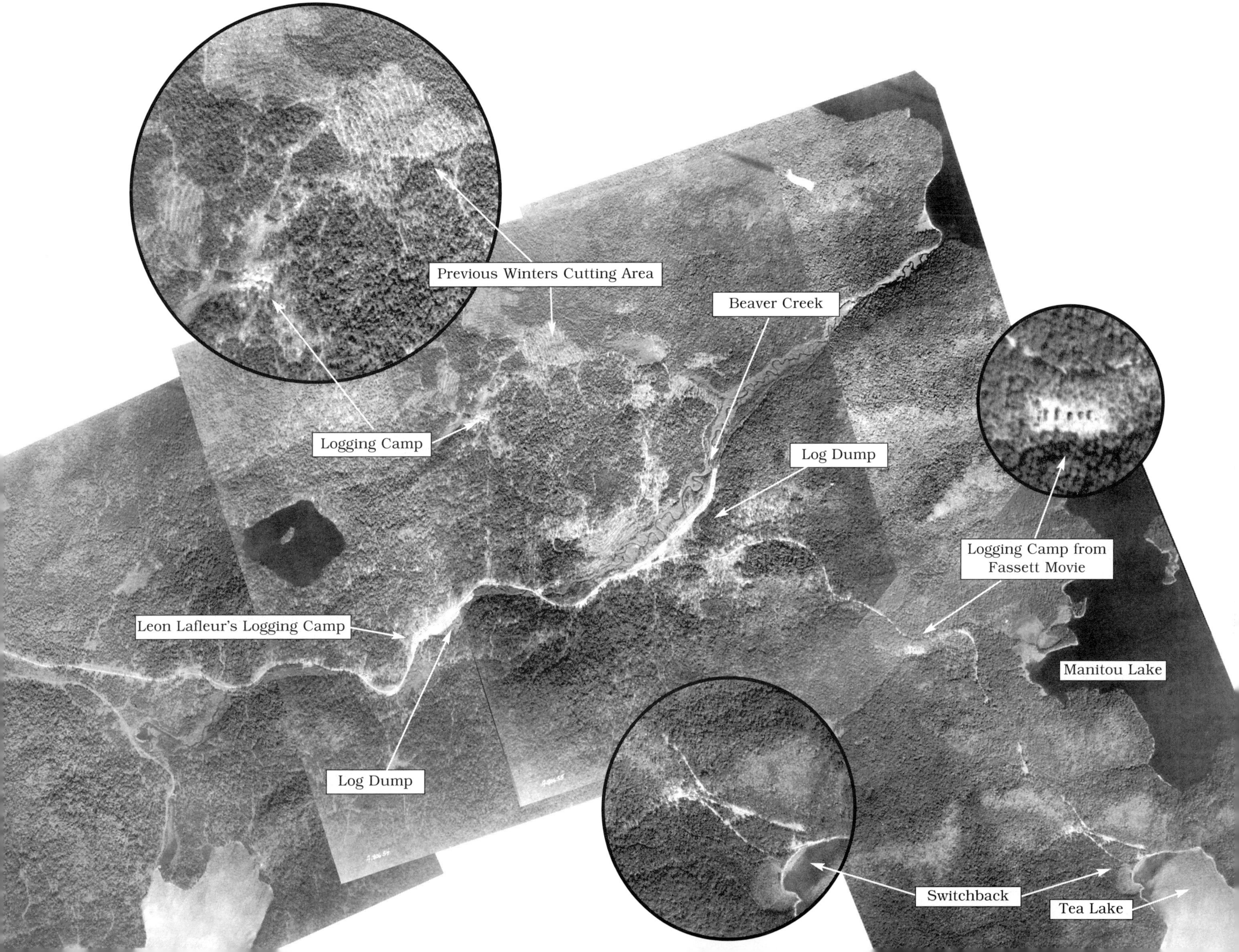

Previous Winters Cutting Area
Beaver Creek
Logging Camp
Log Dump
Logging Camp from
Fassett Movie
Leon Lafleur's Logging Camp
Manitou Lake
Log Dump
Switchback
Tea Lake

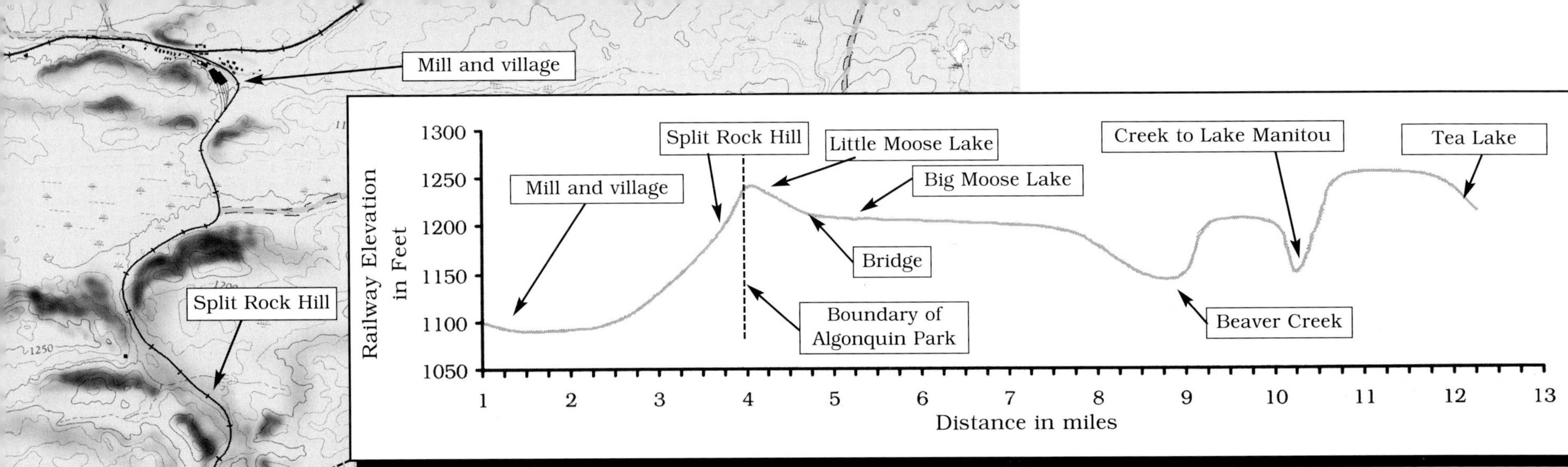

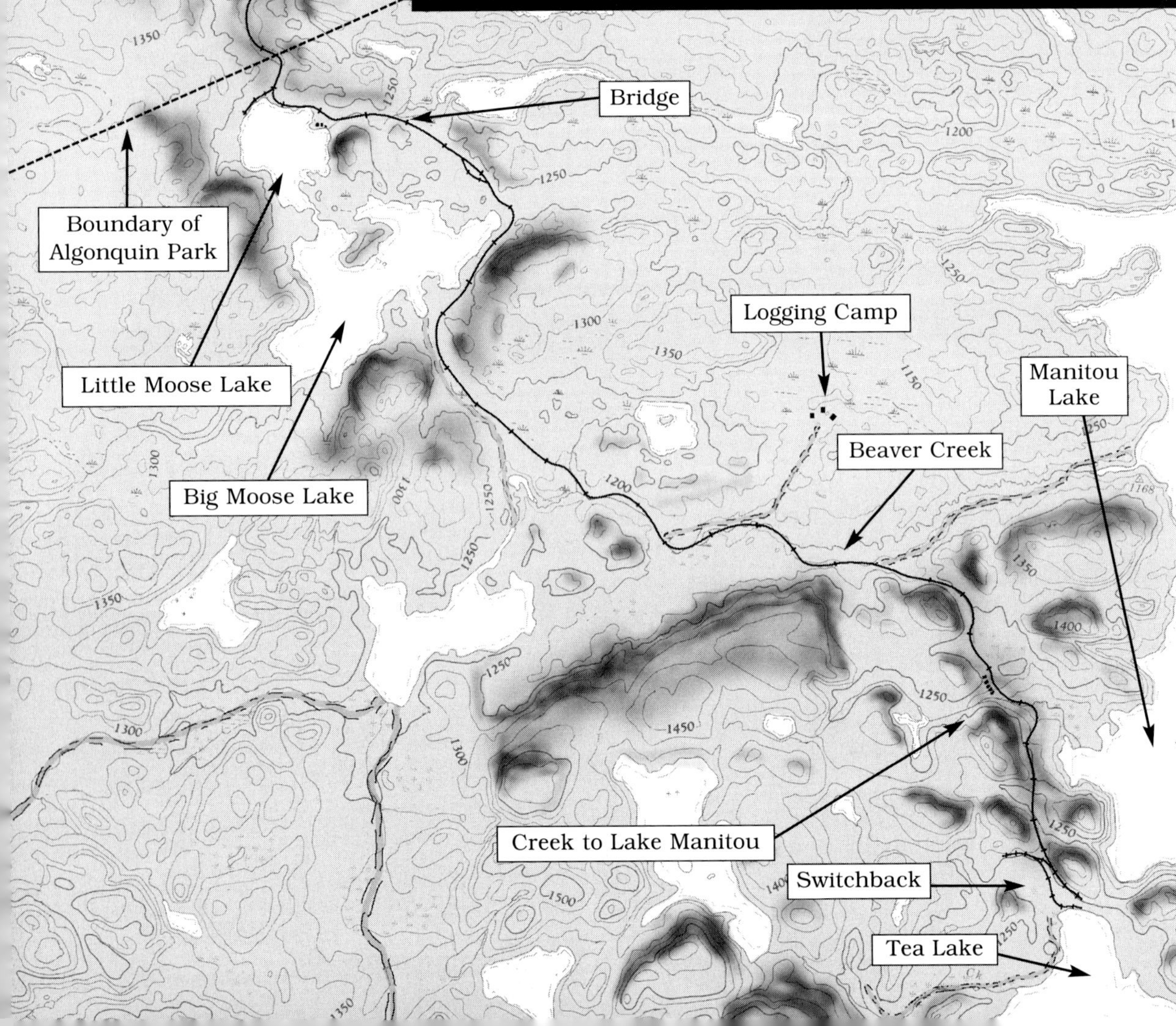

Park. The Fossmill railway crossed two miles (3 km) of swamp, climbed 200 feet (70 m) over the next two miles, and swerved around many hills, before reaching its highest point close to Little Moose Lake, just inside the boundary of the Park. From there on the terrain was rugged highland, dotted with lakes and creeks.

"When they made the line in there they didn't have cranes or anything to remove rocks or make rock cuts," recalls Roger Bergeron. "The old track used to wind in and out and follow the best path it could."

In February, 1924, the surveyors, Thompson and Angus worked with Jack McGibbon and laid out the line. Finnish workers from Sprucedale, Ontario cut a swath 14 feet wide (4 m) through the forest and laid 3,000 ties per mile (1.6 km). These ties were cut on site from whatever unmerchantable timber was available. They then laid and spiked rails to the ties. A team of 30 men laid a mile of track per month at a cost of $23,250 for the first five miles (8 km).

Large logs, supported by submerged logs,

bridged creeks and streams. Max Smith remembers taking teams of horses around these bridges into the camps. "When we walked in with the horses there were a lot of creeks and rivers to cross and there was space between the ties. We used to have to go down through the water with the horses and around the bridges because the horses wouldn't walk over the ties."

As logging progressed, more track was laid. By the winter of 1928, a total of 12 miles (20 km) had been completed. The last section ran east along Beaver Creek toward Lake Manitou.

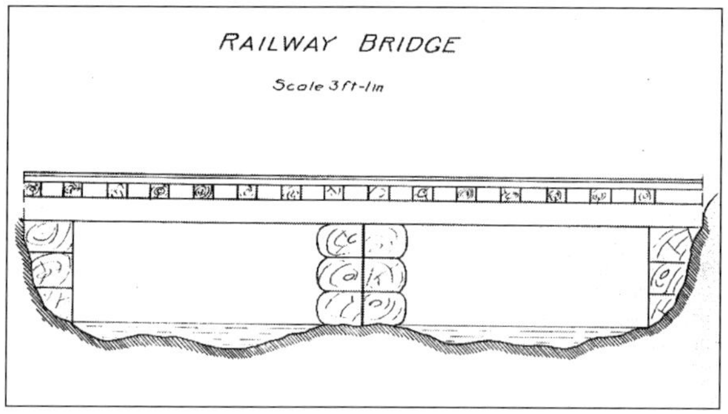

Bridge Plan: *Typical bridge design on the Fossmill Railway.* [UTA]

No 54: *Locomotive steams-up by the lumberyard on a cold winter day.* [UTA]

In the summer of 1929 all the timber in this area had been removed, so the rails on this section were lifted and the line was rerouted south to Tea Lake, adding 4.5 miles (7 km) that summer. At Tea Lake a **switchback** was constructed to allow the railway to descend the hill to the lake-side log dump.

The railway surveyors stayed in a special company cabin called the "Owl's Nest". It was 10 feet by 16 feet by 8 feet high and was framed with heavy timbers. When workers completed a section of track, the steam loader lifted the Owl's Nest onto a flat car by hooking an eyebolt on the cabin roof. The train pulled the flat car to the end of the line where the cabin was unloaded again.

As mentioned earlier, the railway rolling stock was brought from Fassett, Quebec and consisted of two locomotives, two Barnhart log loaders and 60 log cars. A caboose was built when they came to Fossmill.

The company used the "little engine", No. 51, primarily around the mill and lumberyard and areas nearby. It was used for shunting boxcars, sawdust cars, coal cars and log cars into appropriate places. The "big engine", No.54, toiled in the bush.

An early start to the day

Margaret Gleason remembers her dad getting up at five o'clock in the morning. "He'd have his breakfast by himself, then he'd fix up the fires and leave about five-thirty. Before he went he'd always go in and tell Mum that the stoves were okay. He went down to work on his engine, to get it ready."

Allan made his way to the engine house beside the mill. It was a large rectangular

Owl's Nest: *When it was not being used by surveyors it was used by company officials and managers as a cabin for fishing expeditions.* [DS]

building, big enough to accommodate the locomotives and the loaders at night and a machine shop to repair them. The shop was run by two Polish brothers-in-law: Joseph Sliwka, the machinist; and Joseph Kozlowski, the blacksmith.

Although the engine house did not have a turntable, villagers referred to it as the "roundhouse", after the engine houses on the main line railways. John Gleason remembers the times he spent there with his dad. "I used to be down in the roundhouse after school all the time. I saw No. 54 taken apart piece by piece and put back together."

The train crew consisted of an engineer and fireman in the locomotive, and a conductor and two brakemen in the caboose. The caboose provided protection for the train crew in all extremes of weather. The crew's main job was to deliver a train load of 500 to 800 logs to the mill each day in spite of the weather conditions or season. The *Canada Lumberman* reported in 1929 that the railway took out 150,000 to 175,000 saw logs that year.

In the morning, the train crew coupled the empty log cars from the day before to the locomotive and headed into the bush. The locomotive went into the Park in reverse pushing the empty cars ahead of it. When they arrived at the log dump, the loader crews took over the loading of logs onto the empty cars.

Loading logs

The two Barnhart steam loaders were used to load the log cars. The loaders were steam shovels specially adapted for logging by the Marion Steam Shovel Company. For logging, the shovel was replaced with a cable and set of tongs to grab the logs.

John Gleason remembers that the loaders "were powerful little machines. I saw them lift big pine like nothing. They had a bunch of gears, and one big cable went out over the top of the boom, which was fastened right to the

Log dump: *Logs being loaded by the Barnhart loader on the left. On the right, a team of horses delivers more logs to the dump.* [DS]

"Roundhouse": *Caboose, flat car, and loaded log cars outside the Fossmill engine shed. It housed full repair facilities including a machine shop and blacksmith shop.* [CWI]

back of the machine. The boiler and the coal were ballast on the back." John knew the head of the loader team. "Jack Huneault ran one ever since I can remember."

From inside the loader Jack pushed a lever which released a cable and dropped the hook down to the waiting senders, who then hooked it to the centre of a log. Jack braced himself before pulling back on the lever to engage the cable and lift the log. The loader rocked forward from the weight, and when he engaged the chain that turned it, the loader rocked to the side. Through years of experience, Jack and his crew stayed well away from the moving chains, drums, and gears.

As Jack lowered the log toward the waiting log car, the toploader helped him direct it to the best position on the car. As with the sleighs in the bush, the largest logs were placed first, in order to make the widest load possible. When the load was almost complete, **binding chains** were placed around the logs, and the last one or two logs were placed firmly over the chains locking the logs below in place.

The loader moved along rails secured to the top of each log car. When the log car in front of it was full, the loader, by means of a chain, pulled itself onto the empty log car behind it, allowing the car it was previously on to be loaded. Moving from car to car in this way the whole train was loaded, ready for the difficult trip to Fossmill. Most of the time the loaders were left at the log dumps and only the loaded cars were picked up by the train and taken to Fossmill.

***Railway operations**: Scenes from the Fassett logging movie*

1. Partially loaded train

2. Sender hooks a log

3. Loader lifts a log

4. Toploader "directs" the log

5. Top logs sit on binding chains

6. On to Fossmill

Busy day: *Locomotive No. 51 is called in to haul logs at switchback at Tea Lake. No. 54 waits on another branch of the switchback.* [RD]

Heading home

Brakeman watches the load

Homemade caboose

The trip back

Steep hills were the main problem faced by the train crew on the trip back to Fossmill. The Algonquin Park Superintendent's Report in 1930 stated that Fassett operated "a private logging railroad on which there are many heavy grades." The wife of Wellington Pilgrim, an early fireman, said, "All I ever heard him say was how he had to stoke up so there was enough steam to get up the hills."

Even the powerful Shay-geared locomotives had trouble pulling a full load of log cars up the heavy grades. The train crew had to separate the train into sections of as few as two cars before hauling them to a siding at the top of a grade. They repeated this until all the cars were gathered at the top and the train could continue.

A fully loaded train sometimes consisted of 22 cars. This meant numerous trips up and down the hills. John Gleason remembers that the trip "used to take up much of the day. It was a lot of work; they never stopped; they kept going."

A load of prime hardwood

Winter conditions

In the winter, snow and freezing temperatures were other problems to overcome. The train crew cleared the snow with a plow attached to the front or back of the train. The average annual snow fall was 30 inches (76 cm), but the *Canada Lumberman* reported there had been 56 inches (142 cm) in 1926 and quoted railway superintendent Morris Donahue. "The winter before last was the worst we experienced in years," He said. "We were out at 4 a.m. with the plow on an average of three mornings a week to get clear before the days run could be sure of getting through. Sometimes the drifts would be too heavy for us. In that case it was all hands out with a shovel for an hour of digging before we could get moving again. But we always got through."

Sometimes conditions were so bad that Allan Gleason was unable to take the train out at all. His son remembers that there were "times when it was too cold to run. My dad used to park the locomotive up beside the house and let it sit there all day. They were afraid of the rails breaking—60 below zero. They wouldn't let a horse out or a man out working either."

Riding on the railway

Riding on the Fossmill logging railway was a memorable experience for many who had the opportunity. Dr. Duncan MacLulich, a biologist who did a lot of fieldwork in Algonquin Park in the 1930s, accessed the Park using the Fossmill railway. He remembered the experience in a 1976

Snow plow: *Villagers pose on the railway snow plow attached to the train. A link and pin coupler juts up from the flukes.* [DS]

interview.[10] "The sawmill company gave me a ride on their logging railway…that was an exciting ride. It had an articulated engine and on some of the hills they would break the train up and take just two to four cars at a time up that part and then keep going back for the rest. It was terribly rough riding. In the caboose you would brace yourself and hang on and you'd still get bounced around."

Another lucky guest rider was young John Rouselle. One of his early memories is the excitement of rides in the cab of the locomotive. "I stayed with the engineer, I remember going around curves, oh, it was nice—noisy, and it was rough track too." Stella McGibbon said, "It was quite a rough, rickety railroad as railroads go."

On to Fossmill

Split Rock Hill was just outside the boundary of Algonquin Park. From there the railway went steeply downhill, and thereafter, was relatively level most of the way to Fossmill. When the train reached the bottom of the hill, Man Martin and Gordon MacDonald worked their way back across the cars to release the brakes. They could then rest for a while in the little homemade caboose. In the winter they warmed themselves by the cast iron stove in preparation for the work ahead at the main log dump just south of the mill.

"They had a siding where they piled the wood along the track," Roger Bergeron recalls. "They'd have miles of it piled there, about 25 or 30 feet high along the railroad." The mill's log pond held only 1,500 logs—a two-day supply—so new logs were dumped in the pond everyday, either directly from the woods or from this stockpile.

"When they came out of the woods they came across a swamp and over the creek and there was a slight grade up the curve," recalls Peggy McGibbon. "We'd be tobogganing up on the hill and see them coming around the curve. We'd say, 'Here comes old 54,' and the sparks would be just flying out."

When the log cars reached the mill they were taken past it toward the village, past the church, school, boarding house, and store and then backed down a siding that led to the log pond where a couple of log cars at a time were placed for unloading.

Miles of logs: *Stockpile of logs piled south of Fossmill.* [CL/UTL]

Unloading logs

The workers in charge of the pond then took over. "Willie" Maloche and his crew released the chains on the log cars one at a time and rolled the logs onto an angled ramp that was level with the bottom of the cars.

Arriving at the mill

***Log pond:** Logs wait to be unloaded into the pond.* [JD]

From there, the logs tumbled into the pond "like pick up sticks." While they worked, Allan Gleason parked the locomotive beside his house and had supper and a break. When the first cars were unloaded, he returned and moved more log cars into place until all the logs were unloaded into the pond.

When weather was bad or equipment failed, the train did not arrive in Fossmill until late in the day. "It didn't matter how late in the evening it was, there was a train load every day," says store clerk, Stanley Anderson. He remembers hearing the train coming in from the bush and dumping logs in the pond as he relaxed in the boarding house long after his work day was through. He says, "We'd hear the logs thumping down long after I'd quit work."

Allan Gleason's children remember that they sometimes did not see their father all week in the wintertime. Margaret says that it was "late when he came home. It was so late at night that we would all be in bed except Mum. There is no way she would keep us up till pretty near midnight. Mother said he always went around to see us all before he went to bed."

In the morning the logs were sorted and fed, as needed, onto the **jackladder** and up into the mill for sawing.

Dumping logs into the hot pond

Inside the mill: *Mill workers take a break to pose for cameraman in the Fassett mill, 1927.* [CWI]

5
THE LUMBER MILL

"It was a very well organized outfit."

The lumber mill was the heart of the Fossmill operation. Normally it operated all year round, except for a period over Christmas, when repairs and maintenance were carried out.

The work cycle in the mill began when the logs were dumped into the pond. The water washed away dirt and loose bark, and softened the wood to make it easier to cut. In the winter the pond was called a **hot pond** because steam from the mill was pumped into it to prevent freezing and to thaw the frozen logs.

At the pond, Willie Meloche's crew directed the logs to the base of the jackladder, which led to the saws on the second floor. The jackladder was equipped with an **endless-flat-link-chain** with spurs that grabbed the logs and pulled them up to the cutting level. At the top of the jackladder the logs were pushed out of a trough by the **kicker** and lined up on a platform called the **log deck** that was just inside the mill. Several logs could be placed there waiting to be loaded, one at a time, onto the **reciprocating shotgun carriage**.

Start of the mill operation: *Willie Meloche's team dumped logs from the log cars onto the ramp on the left. The logs tumbled into the pond and were directed up the jackladder into the mill. [JD]*

Sawing

The carriage was a steam driven platform on wheels that ran on tracks past the bandsaw and carried the logs through the 10 inch bandsaw blade. The logs were held in place by **dogs**, which were spike-like pointed shafts that were driven into the logs by the **dogger**, one of two men who rode on the carriage. The other man was the **setter**, who set the width of each cut as directed by the head sawyer. The head sawyer stood to the side of the carriage where he could see the logs and decide on the best way to cut them up. Using hand signals (see sidebar page 93), because of the noise, the sawyer directed the dogger and setter as to how to position the cut. He stepped on a pedal that engaged the carriage and it shot forward pushing the log into the bandsaw before shooting back for another cut. Some logs were rotated by the **steam log turner** [11] to make them square before they were cut into planks.

Jackladder

Processing

A number of critical decisions had to be made once the logs were cut into planks. "It was a very well-organized outfit," recalls Max Smith, who still lives near Fossmill. "We were eager to see the operation so they took us through. You would see the lumber coming off these big rollers and there was a big track running right down the centre of the whole

Hot pond: *A dock ran out into the centre of the pond to provide access to the floating logs.* *[UTA]*

Shotgun carriage

Dogs and log turner

Making lumber

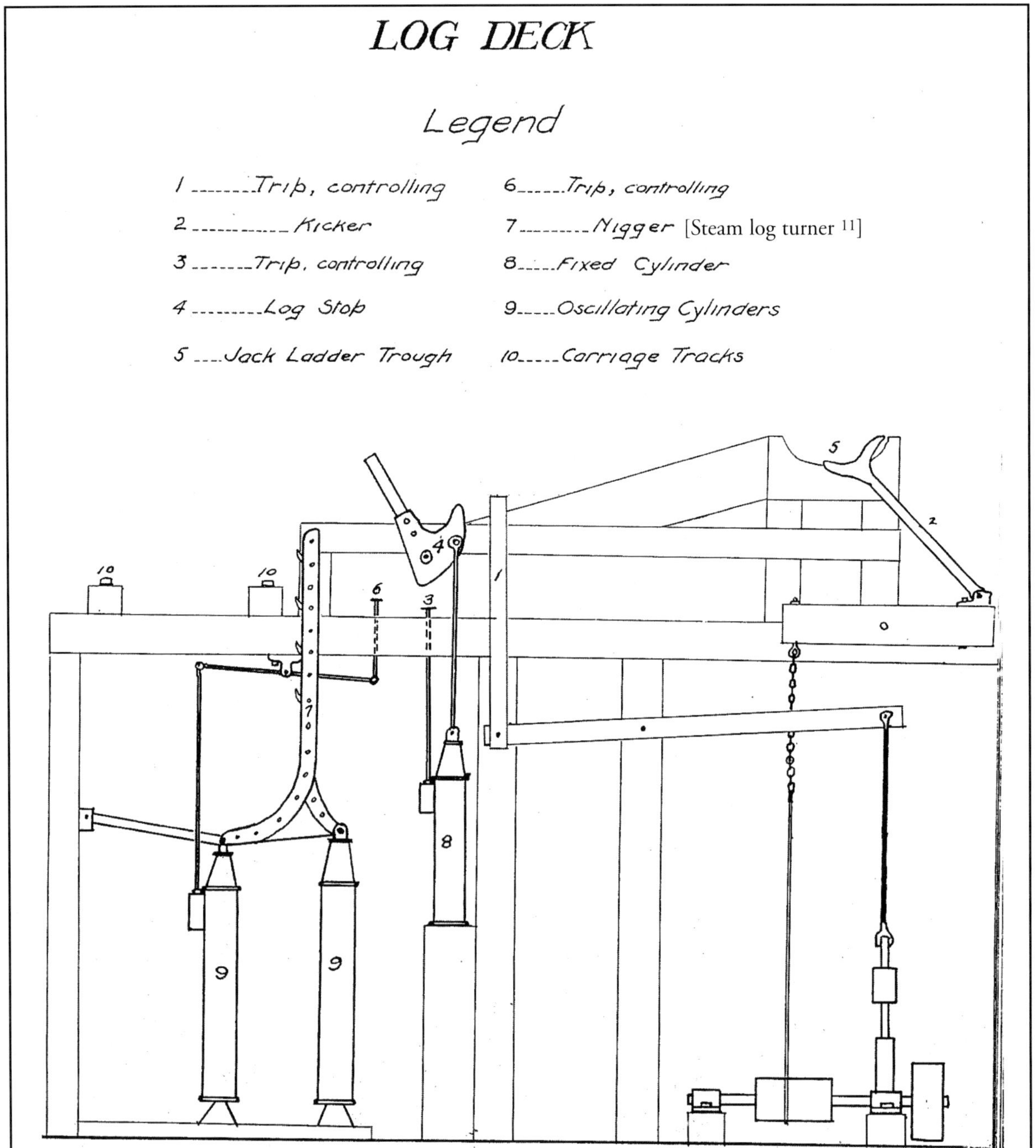

Log Deck: *Cross section of the machinery of the South River Standard Chemical Co. lumber mill.* [UTA]

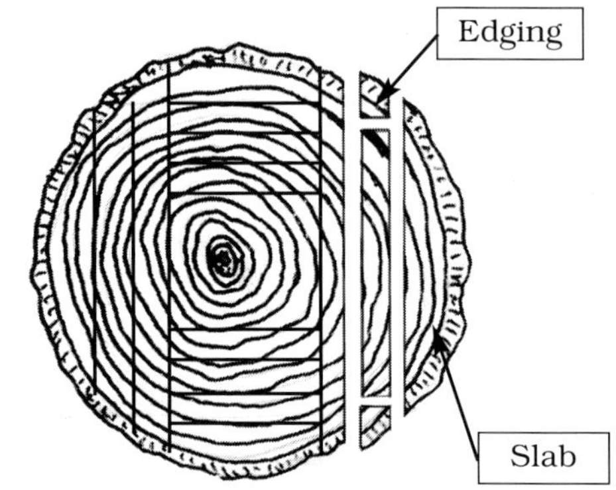

Plain-sawn log

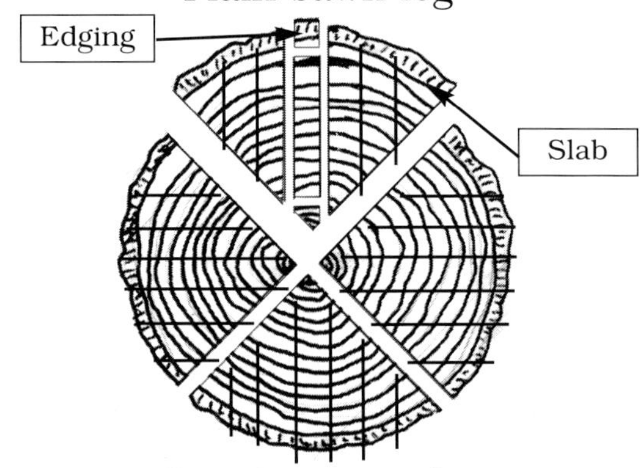

Quarter-sawn log

Cutting Methods

The two main methods used to cut hardwood logs into lumber were Plain-sawn and Quarter-sawn. Plain-sawn was more common because it was easier and cheaper, however the lumber was more prone to problems like warping and bending than the Quarter-sawn method where the parallel grain produced a more stable, defect free and even grained lumber for sensitive and quality applications. For example, the sound board of an acoustic guitar or piano are generally made from quarter-sawn lumber.

thing. It went way out to the railroad track."

Transfer chains and **live rollers** grabbed the planks and slabs, drawing them to various locations. Some lumber was custom cut to various thicknesses on the **resaw bandsaw**. An order might also be planed smooth and to an exact thickness by the planer.

Hand Signs

"Though not a word is spoken Dugas and the two carriage riders work together perfectly, transmitting their wishes entirely by hand signs. For example, Dugas shows his closed fist and the man on the carriage moves the log for a cut six inches thick: a fist shaken means a 12-inch cut and a fist thrown from the body means a 16-inch cut. The motion of Dugas' open hand downward means to loosen the clamps on the log so it can be turned over."

From "A day with a pine sawyer" by Russell Downs from the May 1936 *Timberman.* This article was reproduced in an article on the Cascade Lumber Company in *Timbertimes* written by American logging historian Steven R. Gatke

Shrinkage Allowance

As lumber dried out it shrank considerably and the sawer had to adjust the cut in order to end up with a correctly sized piece of dry lumber after a year of drying. The Student Report on the Standard Chemical Company reported that the sawer there cut the planks:

1 1/16 inches thick for 1 inch lumber
2 1/16 inches thick for 2 inch lumber
3 1/8 inches thick for 3 inch lumber

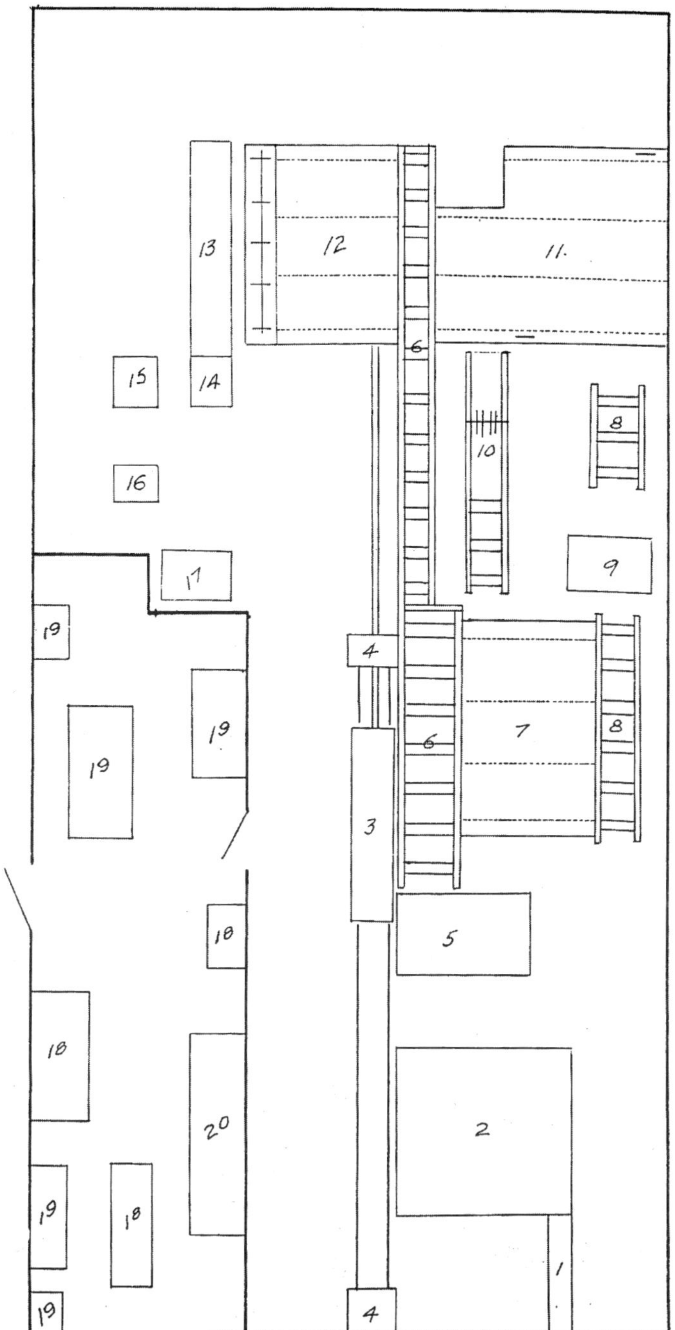

Typical sawmill floor plan: *This plan is of the Standard Chemical Company lumber mill in South River. In Fossmill, the saw filing room was a loft built above the main cutting floor.* [UTA]

Sawfiler: Paddy Dwyer

Sawfiler's loft

Trimmermen

A floor plan of the Fassett mill is not available, but a plan of the similar Standard Chemical Company mill in South River shows the complexity of the technology (see page 93).

In the Fassett mill, saw blades were sharpened in the third floor **sawfiler's loft**. The bandsaw blade was sometimes replaced with a freshly sharpened one during a 10 hour shift and always at the end of the shift. The in-shift change often gave the men their only break.

Once the lumber was cut to width in the **edger**, and trimmed to length, it went into the sorting shed, a long open wing off the main building, where it was separated according to species, thickness, and **grade**.

"It was the grade of the lumber that was worth the money. The lumber you see today, they would just grind that up," recalls John Gleason. The **grader** determined the quality of the lumber and marked each piece accordingly with chalk (see opposite). The **sorters** pulled the boards off the moving belt and put them onto the appropriate tram cars. When they were loaded they were pushed to the lumberyard or to special storage sheds. John Gleason remembers the special sheds. "There was a difference in the way they piled it too. You take the curly birch and the bird's eye maple, there were all these sheds and it was all piled on end." (see photo page 96)

The lumberyard

Most of the lumber, however, went to a huge yard east of the sawmill. This lumberyard was arranged for the efficient drying, storage, and shipping of the lumber. Numerous tracks came out of the sorting shed but only six tramways went into the lumberyard. Trams were pushed onto a sliding platform called a **transfer track** that moved at right angles to the tramways. This track was aligned with the track on the appropriate tramway and trams

Grading Lumber

Graders had to make quick decisions as to the quality of lumber entering the sorting shed. The requirements for each grade were complicated and graders had to have an eye for the appropriate grade of each piece. Following is a simplified explanation of the basic requirements.

FAS (Firsts & Seconds): The top quality lumber had to be at least 6 inches wide and 8 feet long and had to be 85 to 90 percent defect free on the poor side.

Select: The second quality had to be at least 4 inches wide and 6 feet long and 85 to 90 percent defect free on the best side and 67 percent free on the poor side.

No. 1 Common: This grade had to be 67 percent free of defects and produce 4 to 6 inch wide and 2 to 3 feet long defect free pieces.

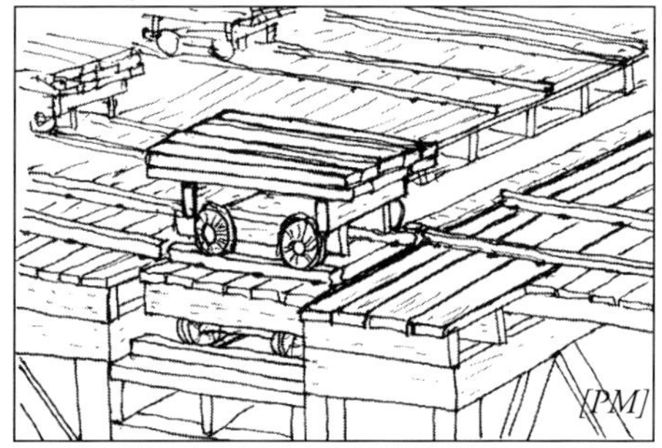

Transfer track: *A gap in the platform allowed a moving table to transfer lumber trams from mill to the tramways. [RD]*

Tramway: *Six tramways ran into the lumberyard. [CWI]*

To the yard

were pushed off the transfer track onto the tramway and out into the yard (see above).

The tramways extended 1000 to 1500 feet (300 to 460 m) into the lumberyard. They had just enough downslope for gravity to assist the men pushing the loaded trams to the appropriate lumber piles. Donat Huneault remembers that "One time the damn tramway broke and a guy went down with a load of lumber—he didn't get hurt though."

The yard team, coordinated by foreman Hilaire Bedard, unloaded the trams. The tramways were up to 20 feet (6 m) above the ground. The yard crew piled lumber from the ground up to the tramway and up to 20 feet above, making some piles 40 feet (12 m) high.

When there was a surplus of one particular type of lumber, a second pile was built behind the first. A **roller** was laid on the first pile and the boards were pushed back on it to the second pile, and both piles progressed upwards together.

The pilers wore heavy protective leather aprons and gloves. They were skilled at keeping the piles aligned and placing strips of wood called **stickers** between each layer of lumber to allow air to circulate and dry the wood. "Those guys were expert pilers. You could just look at the front of the pile and it was as straight as a wall," says Donald Staniforth.

In order to load the sold lumber into boxcars, the company built five railway spurlines between the lumber piles. The CNR supplied boxcars to ship the lumber on its railway. Locomotive No. 51 pulled or pushed the boxcars into place in the lumberyard ready to be loaded. Shipping crews piled the orders, piece by piece, into the boxcars.

Each piece was individually tallied as it was placed in the boxcar. Louis Demers remembers his first job for the company. "When I was young I used to stamp and put a number on every piece of lumber that went into the cars." Loaded boxcars were **shunted** to the railway siding, where they were picked up by the CNR for shipment.

Lumber spaced for air drying

Sidelines

All waste from the lumber production was put to use and actually generated around 9% of the company's income (see graph page 114). Stanley Anderson described some of these sideline operations. "One of the things they did have, that none of the small mills had, was the **lath mill**. The lath machine also made **pickets**, four feet long, for the furniture trade, to make legs for tables and chairs. The pieces were put into bundles. The big heavy slabs of maple were used for these items in order to reduce the waste. It also made it profitable to take out a crooked log. A crooked log always had a heavy slab."

Mill workers Donat Huneault and Louis Demers described the jobs they did as teenagers. Donat recalls, "I worked in what they called the picket mill. We used to get orders and I cut the pieces to length." Louis says, "I did all kinds of jobs there when I was young. I worked with slabs. I put them in carts so they could carry them away. A horse came and they would empty the box and then I would fill it up again. I worked making pickets too. I worked in the yard piling those pickets, all kinds and shapes. I took them to the yard with the horse and unloaded them into different drying sheds."

Slabs not suitable for pickets or lath were cut into stovewood or cordwood and sold to people in the village or nearby community. Wood unsuitable for these uses was cut into short pieces for easier handling and chewed up in a machine, appropriately called the "**hog**", in preparation for use in the furnace where it was burned to produce steam.

The tanbark from the hemlock trees and the excess sawdust and shavings were sold to the Anglo-Canadian Leather Company in Huntsville. It was shipped in open **gondola cars**, which were filled from a spout beside the boiler room.

Cutting slabs into pickets

Cutting pickets: *Donat Huneault at age 17 cuts pickets to length.(see page 197 and 212 for photos of Donat when he was older.)*

Special sheds: *lumber sensitive to staining was stored vertically. [DL]*

Tragedy

On August 1, 1927, Charles McFaul, the stationary engineer in charge of the steam generation plant, noticed sawdust piling up in the mill and decided to transfer some of it into waiting gondola cars. When he went outside, the car under the spout was full. There was no one around to move the car so he decided, since he had seen it done many times, that he would move another car in place himself.

"There would be maybe half a dozen cars in a row. When a car was filled they were just edged along," Peggy McGibbon recalls. The cars were on a grade and by releasing the hand

Loading boxcars

Checking that the order is complete

Lumber pilers: *Chisholm's Albert Atkinson (left), and Joe Kunkel (centre) working in the lumberyard 1929.* [CWI]

Sawmill daily wage rates Standard Chemical Co. Summer 1926

Typical daily wages paid in the 1920s for a 10-hour day.

Dogger	*$3.25 to 3.50*
Canter	*2.75*
Filer	*10.00*
Sawyer	*7.50*
Setter	*4.50*
Trail Sawyer	*3.25 to 3.40*
Edger	*4.50*
Slash Table	*2.50 to 2.75*
Trail Edger	*3.00*
Trimmer	*2.50 to 2.75*
Culler	*3.50 to 3.60*
Boardway	*3.00*
Greaser	*3.00 to 3.40*
Jack Ladder	*2.75 to 3.00*
Lake Men	*2.75 to 3.25*
Watchman	*3.25*
Engineer	*4.00*
Lathmill Man	*3.00*
Lathmill Boys	*2.25*
Jackers	*3.00*
Pilers	*3.25*
Slab Wagon	*3.00*
Yard Teamster	*3.00*
Shippers	*3.00 to 3.25*
General Yardman	*2.75 to 3.00*
Feeding Resaw	*3.25*
Trailing Resaw	*2.75 to 3.00*

brakes the cars rolled forward slowly, putting the next car under the spout. "On the gondola car they had individual brakes. You went up a ladder and in the centre of it there was a big wheel and you spun the wheel to put the brakes on."

Charles McFaul must have released the brake and "not knowing the proper procedure he leaned in the wrong direction when he applied the brakes again." When the brakes engaged McFaul was thrown from the car.

Feeding the firebox

John Orr oiling the flywheel bearings

"He fell off. The car didn't stop, and he was killed," Peggy recalls.

Power plant

The power for the mill was generated by burning sawdust and scrap in the firebox of the mill furnace to produce steam. The steam drove an engine that turned a huge **flywheel**. Max Smith remembers that "The wheel that drove the whole thing was about 4 feet thick and 14 feet in diameter, and it turned slowly all the time. A belt went around that wheel and was hooked onto a bunch of small pulleys; those pulleys went like the dickens and ran the whole mill."

Peggy McGibbon recalls that the steam engine also ran a "huge dynamo that gave hydro to the office, the store, and the special boarding house that was above the store." It also provided power for lights in the mill and the managers' houses.

Lumber mills driven by hundreds of belts turned by a steam engine were quite intimidating. One visitor commented that the area "was a big ominous place and very noisy." In the dark of the lower levels of these mills, flashes of blue-flamed static electricity jumped from the belts to the bodies of people who went too close.[12]

Peggy McGibbon recalls visits to the mill's boiler room and glimpses of the firebox. "Underneath the mill there was this big 'inferno' I called it. The floor was all heavy metal and there was a big disc that swung aside and you could look down—I tell you it was the fires of hell. That's what I think of every time I hear the word 'inferno'." One man remembered the firebox was an ideal place to get rid of an unwanted litter of kittens.

Wages and working conditions

Mill workers put in 10-hour days, 6 days a week. Sunday was the only day off and there were no paid holidays. They worked from 7 a.m. to 6 p.m. with a one-hour lunch break. One worker recalls that a spare man replaced him during lunch.

One woman remembers that her father came home for lunch. "He ate a big meal. After he got through he sat in his rocking chair and slept for about ten minutes, and then he went back to work. That was his daily routine."

During the 1920s, wages ranged from $2.50 to $3.50 per day for most workers. Specialists like the engineer and the sawfiler made $4.50 and $10.00 respectively. Any money owed to the company store was deducted. The company also paid workers' compensation premiums to the government.

There was some union activity in the Ontario logging industry in the 1920s. Most of it, however, was further north than Fossmill and was concentrated among pulp workers. One resident commented, "I don't think anyone tried to organize a union at Fossmill, even though nobody had holidays or pensions or nothing." It would take a new generation of workers and a world war before the workers would turn to unions for improvements.

Over-production

Lumber production at Fossmill increased rapidly through the early years. In 1926, they cut 8 million board feet and in 1928, they cut over 12 million board feet. Sales, however, lagged well behind production. In 1928 only 8 million board feet were sold.

During the boom years of the 1920s, the practice of business was to build huge inventories in anticipation of future sales. Fassett had invested heavily in the railway and mill and wanted to use them to capacity, so it based its cut on the supply of logs, rather than on expected sales. In 1928, the *Canada Lumberman* quoted the company as saying, "We ourselves, are planning to take out a larger cut than last year which is not due to the market, or any anticipation of the market, other than that the territory we are covering will give us a larger quantity of hardwood this year than last."

New workers were hired to expand production, and the village of Fossmill expanded. The original families from Fassett and Chisholm welcomed their new neighbours and the community grew.

Warning!

In an article in the *Canada Lumberman* in 1929 called "Hardwood Logging in Canada" N. C. Jones warned that the cost of building a railway "is apt to be too great a burden...It is natural having installed the line to strip the limits, taking small trees and utilizing as much as possible," and that this "system leads quite naturally to the destruction of the hardwood forest."

Overflowing: *View of the Fossmill lumberyard from the south. Over 14 million board feet of prime lumber waiting to be sold.* [CWI]

Home sweet home: *A typical house in the main residential area of Fossmill.* [CWI]

6
THE VILLAGE

"...a strange little settlement."

It was a hot Dominion Day, 1927, when Rachel, Claire, and Octave Denis arrived in Fossmill from Montebello, Quebec. Their father, 55-year-old Albert Denis, had been hired the year before to work as a carpenter in Fossmill. He had completed building his house and his children were now joining him.

Rachel, age 15, Claire, age 10, and Octave, age 8 were three of his six children. Life had been difficult for them since the death of their mother in 1923. Their older sisters, Laurette and Juliette, had raised them until Rachel had been sent to a convent. Claire and Octave had been sent to an orphanage in Montebello, and their little sister Cecile had been adopted.

Claire, in particular, felt relief that day. In a 1998 interview, she said that she and Octave had been malnourished and treated badly in the orphanage. Fossmill was a chance at a better life.

Rachel returned to Fossmill to "keep house" for her father, brother, and sister. At 15, she became a substitute mother to her family, being responsible for all the family's cooking, washing, and house cleaning.

"There were few luxuries

Looking west: *The main residential area of the village as seen from the sawfiler's loft in the mill. The white building in front is the stable and the Church is under construction centre-right.* [PL]

and little money at the time," notes a biography [13] of the Denis family. "In addition to the responsibility of carrying out family duties, life for a 15-year-old French-speaking girl from Quebec, presented difficult challenges in this new English speaking community." The main outlet for Rachel's French-Canadian culture was the church, which her father had helped build.

Albert Denis had built their house in the centre of the village, between the church and the school, on a spot on the hillside beside the houses of the Desjardin family and the family of Bill Casselman, the CNR station agent. It was a small, flat-roofed bungalow about 12 feet by 20 feet. The front of the house was at road level and the back was supported five feet above the ground. The location was not ideal. Claire recalls that the house was "in the gully with all the clay and corruption."

Denis' house: *Albert Denis built the little bungalow seen on the left, the church is on the right and Tom Howard's house is in the background. [LN]*

Exploring the village

Because the Denis children arrived early in the summer, they had two months to explore the village and area before school began. The main residential area was arranged in two neat rows west of the school and church. East of the church, beside the mill and lumberyard, was a second residential area. There the houses were arranged more haphazardly along a trail that wandered along the side and top of a small rise beside the company railway track and into the bush east of the village (see drawing page 104).

Although many workers built their own homes, they did not own them. They were built with cull lumber from the mill, and were on company property. Residents paid $5 to $10 a month rent. "If you didn't work for the company, you couldn't stay in the houses. You had to vacate, that was it," remembers Roger Bergeron. The houses were made of unpainted wood and did not have foundations. Margaret Gleason recalls that her house "wasn't insulated or anything. The walls were just boards with felt paper over them, and then wallpaper over that, plus wallpaper over that, plus wallpaper over that, every couple of years." Stella McGibbon said, "they were all just lumber houses, put up in a hurry."

The layout and amenities in each house varied widely depending on when it was built, where it was built, who built it, and the family's income. The houses beside the lumberyard were built later than the others and were not as desirable.

By the lumberyard

"They were shacks, they were all shacks! Ours was probably one of the better ones," says Roger Bergeron, referring to the house he moved into in 1928 when his family came to Fossmill. "It was one storey. It was a fairly big house with a big porch on the front facing the lumberyard, the coal bin, and the lumber track. There was no indoor

Relaxing: *Herve Beauchamp and boarder Jimmy Bellefeuille on the front porch of the Bergeron's first house in Fossmill.(the large house centre-left in the photo below.)* [RB]

bathroom—nobody had one—nobody had running water; they had pumps outside in a well. There was no electric lights, and all we had in the house for heat was a cook stove."

Like the rest of the houses, Roger's house was not insulated, and the walls were boards covered with cardboard and wallpaper. With only one stove, winters were particularly trying. "I remember, in the middle of the winter, getting up in the morning and finding the water frozen on top of the pail by the sink. I remember the linoleum all curled up. The wind used to whistle right through those houses. Every winter the house would twist and move—POW! you'd hear it in the middle of the night.

"The only person who had a chamber pot was my mother," says Roger. "Everyone else went outside. You made damn sure that before you went to bed at night you took care of your business. It wasn't funny to get up in the middle of the night and trudge down to the outhouse."

John Rouselle's house was also typical of these houses. "Our house must have been 20 by 20 and the kitchen and the living quarters took up half of that. The other half was separated into three rooms, with curtains. My dad used to sleep with my mom in their room, the boys would sleep in the second room, and the girls in the third." Another villager also remembers curtain walls. "The house was about 24 by 24 with one room for my mother and dad, and another big room separated by curtains on a wire running across. You would pull the curtains for privacy."

East end: *The row of houses beside the lumberyard.* [CWI]

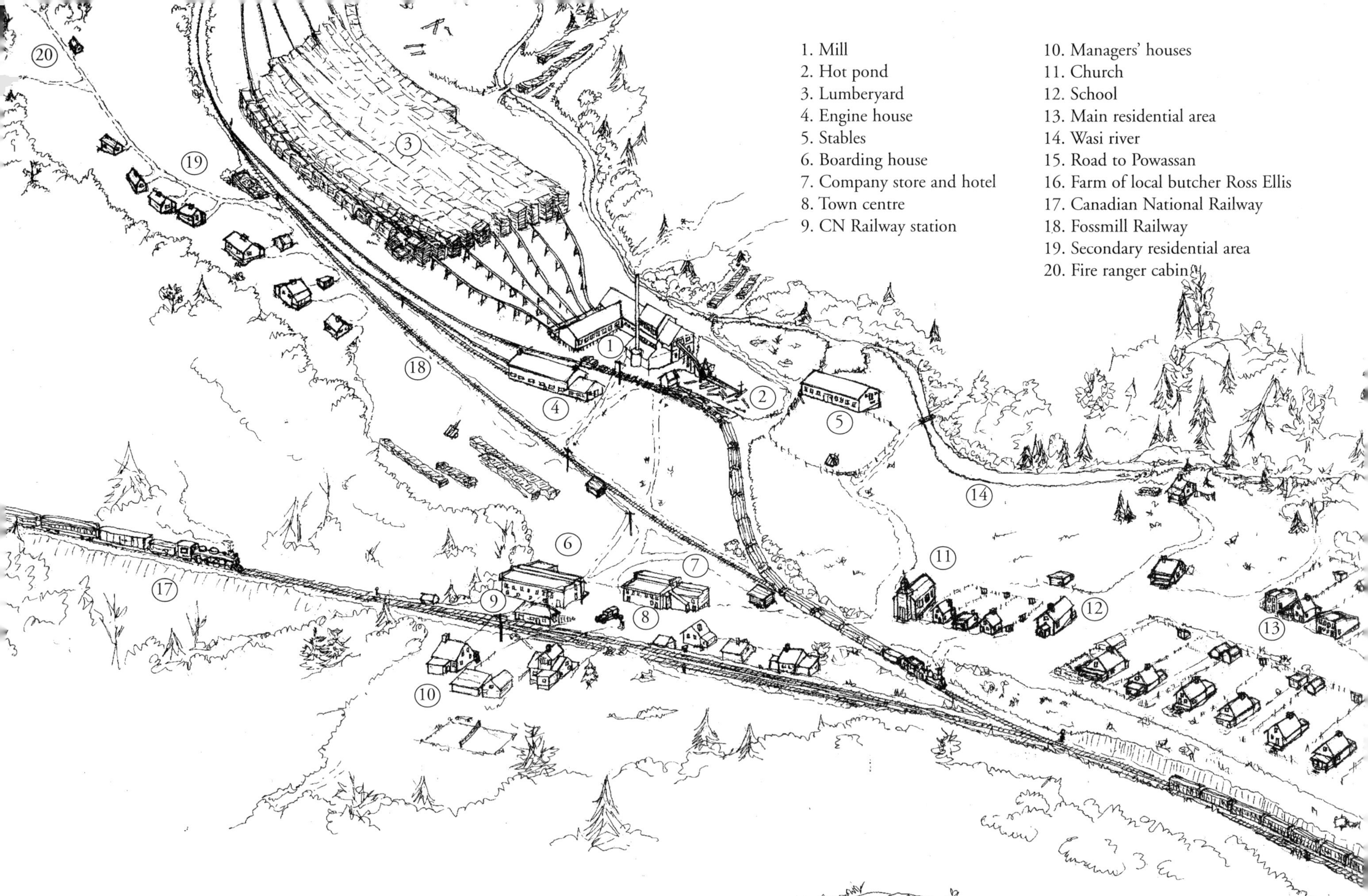

Fossmill, Ontario and the Fassett Lumber Corp. 1924-1934

Main residential area

Most of the houses in the main residential area had been built in the early days and were a little more substantial. The Gleason's house was typical of these houses. It was the first house west of the school.

Margaret recalls that in the back of the house they "had a big kitchen and dining room combined, with a big cook stove in it. Off of that, we had a little pantry where all the cupboards were. We had a little sink and a wood box. The toilet was outside." There was an archway and stairs between the kitchen in back and the living room in front. She says they had "a good-sized living room with a box wood-stove—Dad would keep the fire going all night. Off of the living room was my parents' bedroom, and off the kitchen/dining room there was a bedroom—now, the bedrooms weren't very big—and we had two big rooms upstairs."

These houses were also without electricity. Lighting was provided by coal oil lamps or in later years Aladdin lamps with **mantles**. Margaret recalled a special lamp that hung in the kitchen. "It had a real beautiful shade on it. It was just like those stained-glass lamps with scenery on them. It was beautiful. Mother thought the world of this shade, so she didn't let us light the lamp. I think I only lit that lamp once in my life."

Gardens

A special feature of the Gleason house was the garden. Because the house was on a corner, the garden wrapped around two sides of it. Grace Gleason was an avid gardener. Her

Housing: *Morris Donahue's house on Fossmill's main residential street. [LN]*

Gleason's house: *One of the earliest houses built. It was known for its flower garden.* [MF]

son, John, recalls that the garden was filled with "all her flowers. You could see them from way downtown. We used to travel all over the country. She'd get flowers here and there and exchange her flowers with others. She had a lot of flowers." Other villagers also did gardening for pure pleasure. John Orr was the stationary engineer who replaced Charles McFaul after he was killed in the gondola accident. John's son Earl says, "Dad always had a big garden at the back. He was a great gardener. He grew all kinds of vegetables but his pride and joy were his flowers."

Most families planted vegetable gardens and raised animals for food. One resident says, "We had a garden full of vegetables, and we had chickens. We used to get a pig in and feed it the scraps." Another says, "Out back there were sheds for the animals and fenced in areas."

Most houses had root cellars dug into the ground under the house. "In the fall, when we harvested the crops, we put them in the cellar and covered them with earth," recalls one villager. "We had lots of preserves, pickles, blueberries, and raspberries."

The company tore down Foster's buildings beside the railway and built a duplex and two single houses at the foot of the CNR embankment across from the company store

The other side of the tracks

The best homes were those reserved for Tom Howard and Jack McGibbon, the company managers. "This was a strange little settlement," recalls Peggy McGibbon. "The railroad went through and everything was on the south side of the railroad except two very good houses, painted white and green, that were on the north side of the track, up on the hill, all by themselves."

The Howard and McGibbon houses were similar. They were two of the few buildings in the village that benefitted from the electricity produced by the mill's generator. "We had electricity from the mill," Stella McGibbon recalled. "It was direct current, so you couldn't use electrical machines. We didn't care about that. We had the light."

Peggy's house

Peggy McGibbon gave a detailed description of the family home in a 1996 interview. "There were three bedrooms and the bathroom upstairs. The bathroom had a nice bathtub, basin and toilet with running water, a medicine cabinet, and a nice cupboard off to the side." Under the eaves there was storage space for her father's hockey equipment and her mother's riding gear.

"We had a stairway with a varnished railing and a gate across it to keep the babies from falling down the stairs. There was a playroom downstairs, across from my parents' bedroom. Us kids had blackboards and toys and stuff in there.

"The front of the house was all living room, two-thirds family room and the last third the parlour." The parlour held a box-stove placed next to her parents' bedroom. The room also contained a piano, an Atwater-Kent radio, a couch, and a "big oak library table with a nice chair. That's where I did my homework. In the family room we had nice wicker chairs, a couch, a Victor Victrola gramophone, and a little stool that was used to reach up and put the needle on the record.

"Outside, across the width of the house was a sunporch. A swing, chairs, and a table were out there. Our part husky, part Saint Bernard, lived on the sunporch in the wintertime.

"At the back of the house there was an addition with double windows on the west end. It had a nicely finished hardwood floor, oak dining room furniture, a buffet, and a round table and chairs. My mother's sewing machine was by the windows. There was a rocking chair and a big wood stove—I mean big! My father loved a big stove." Beside the stove was the kitchen with counters and a sink. She said they "used water from the mill for washing, and a little pump on the sink for drinking water." This pump was connected to a 22-foot-deep well east of the house.

Peggy's mother, Stella, said that her house, like all the houses in the village, was not insulated, but, she said, "I don't remember ever being cold, there was lots of wood. We were very comfortable." When asked who filled the stoves, Peggy replied, "My mother—my father wasn't at home."

The McGibbons did hire housekeepers. Peggy recalls that "Mrs. Short worked for us quite regularly as a housekeeper. When she left she'd say 'there missus, you're clean again' and then we had Elva MacDonald and then Grace, she was the sister-in-law of Mr. Casselman. In 12 years we had five people over half the time."

Bernice Martin (Leyzac), daughter of Brakeman Man Martin, recalls visiting the

Vegetable garden: *Most families grew their own vegetables. "Aunt Maggie" Donahue preparing her garden. [LN]*

McGibbon house and that they "had a typewriter we used to play on. They had this big, big house. To me it was just like a mansion."

Private hotel

The company also built a small private hotel above the store. Roy Summers, the paymaster, and Alton Douglas, the office manager, both lived in Powassan 12 miles (18 km) away, and stayed in the hotel most weeknights, especially in bad weather. The sawfiler and his wife also lived there for a while.

Mrs. McFaul, wife of the stationary engineer killed in the mill accident, and her daughter Agnes, ran the hotel and lived there. The rooms were equipped with "beautiful" furniture, and it was one of the few buildings in the village that had running water, indoor bathrooms, and electricity.

Stanley Anderson recalled that there were five or six rooms for Sydney Staniforth and other company officials and guests to stay in when they visited. Margaret Gleason says, "What we called the 'high muckity-mucks', lived up there."

Sydney Staniforth

Although Peggy McGibbon remembers that "Staniforth came to Fossmill only as a big shot" he was well liked by the villagers. Nora Gleason who knew Sydney and his wife in Fassett, Quebec said that they "weren't stuck up or nothing. They were very nice people. I liked Mr. Staniforth. He'd help any kid. I

Hotel keeper: *Agnes McFaul managed the hotel above the office.* [CWI]

Boarding house: *Mill workers relaxing in front of the boarding house.* [RB]

Village centre: *(Below) The company store and hotel are on the left, and the boarding house is on the right. A new larger CNR station is in the centre and Tom Howard's house is across the railway tracks behind the station.* [RD]

Visit: *Sydney Staniforth and family (right) on a fishing trip at Fossmill, accompanied by Jack McGibbon and daughter Peggy McGibbon, At first Peggy did not recognize her father in this photograph because "My dad didn't dress that fancy." Maybe they "got him to dress up," she said.* [DS]

Living beside the CNR: *Mrs. Martin outside her house beside the main line of the CNR. The Danis' house is behind her. Her son Lionel is on the left and daughter Bernice on the right.* [BL]

remember he gave a prize one year for the best pupil in mathematics and I won it. It was a diamond broach. I lost it in Toronto Union Station. I felt so bad!"

Mill worker Donat Huneault remembers one of Sydney Staniforth's visits to the mill. "I hit old man Staniforth with a piece of wood. I was running the hog and he came around behind me and I got him, BAM!—'I'm sorry,' I said. 'Don't be sorry, I have no business being here,' he said."

Sydney's son Donald recalls that his "dad liked to get out into the woods and walk around and know what the hell was going on." Sydney regularly inquired about the workers and their families, especially those who had worked in Fassett.

Charles Dale

Company President Charles Dale also came to Fossmill occasionally. Nora Gleason remembers that Mr. Dale came up "every once and a while. Money to him was no object. He was that type of a guy—a real swinger! You know—dressed. I was in my teens, and I used to think, for an old man, which he probably wasn't then, the way he dressed, white pants and white shoes—a real, what we'd call, a 'ladies' man'. He was just money, and having a good time. You knew he was in charge, but I don't think he ever put his nose in the bush. He wasn't the type. He drove a big car and sometimes he'd have a chauffeur." Of course, along with this image went rumours of mistresses and villas in Europe.

Which pedal? *A Donahue learns to drive the family car.* [LN]

Automobiles

Across from the hotel the company built a garage. Peggy McGibbon remembered that "Alton Douglas kept his car in there and on the north side of the store they built a platform that had gas pumps on it."

These gas pumps serviced the new automobiles that became popular in the late 1920s. The earliest cars in Fossmill were Tom Howard's and Jack McGibbon's. The McGibbon car was a 1925 Dodge touring car. "That was a beautiful car—navy blue—varnished wood spokes and blue leather upholstery," Peggy recalls (see photo page 48). Jack's wife, Stella, learned to drive and often took her children and their friends on outings. When weather permitted, the doctor, veterinarian and other services arrived by car. Roy Summers and Alton Douglas drove back and forth to their homes in Powassan.

Although expensive, automobiles were not out of reach for many. Medium priced cars cost around $900, and a Model-T Ford sold for around $450, which was more than half a year's pay for a mill worker. Many saved and bought cars outright. Others bought them on credit.

Roger Bergeron's father had an "Overland Whippet with wooden spokes." Roger also recalls a Chisholm man who drove to his job at the mill every day. "Joe Kunkel had an old Model-T that he used to park by the engine house. We used to pass by it going home, and

1929 Overland Whippet: *Mill worker Amedee Laferriere (left) and brother Oscar (centre) show off their 1929 Whippet car.* [GL]

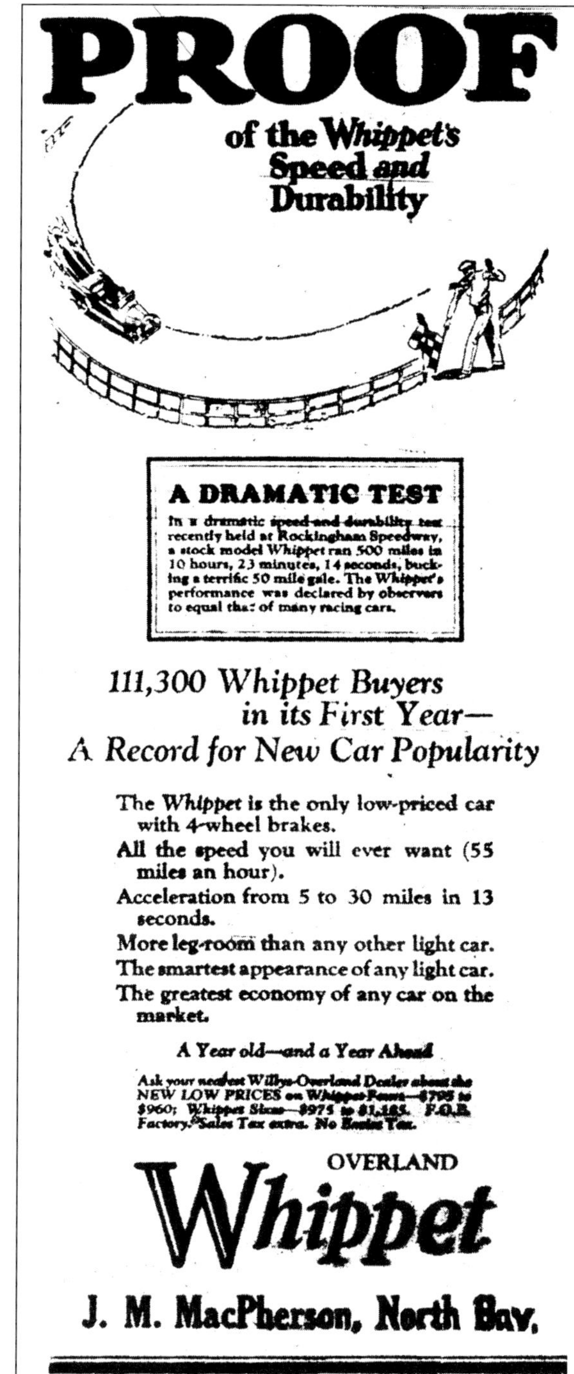

we'd toot the horn."

Villagers remembered cars that would not start or were stuck in ditches. Dirt roads received little maintenance and were vulnerable to rain in the summer. They were often impassable in winter and cars were usually put on blocks during the season. But as roads improved and automobiles became more common, Fossmill became less and less isolated.

By 1929, one out of every two families in Canada owned a car. Earl Orr remembers his family had two 1927 Dodges. One was a touring car with a "roll-down canvas top with windows that flapped." Peggy McGibbon remembers that "In those good times there were at least 30 cars in Fossmill. Those young fellows that were working all bought cars."

Rachel Denis met one of these young men, Amedee Laferriere. Amedee and his brother Oscar had "pooled their money and had some of the nicest cars of the day," states the family biography. So when Amedee "came a-courting…there were enjoyable trips to North Bay, Montebello, and Ottawa."

Peggy McGibbon tells of her father's interest in Canada's first Chrysler Eight. He saw it at the Canadian National Exhibition in Toronto. It was covered with tarpaulins and big signs announcing the new model. "It was only going to be shown when the CNE opened," she says, "but my father bought it before they put it on display. Because he was taking it before they had their parade on Yonge Street, it had the banners that went with it. There was a big wheel on back and flags. For years after that, everywhere we went—Pembroke, Ottawa, Lachute, Montreal—you'd see people gather around that car!"

Railway station and boarding house

The railway station was east of the gas pumps and the road that crossed the tracks to the managers' houses. After the Fassett Lumber Corporation moved to Fossmill, the

Stanley Anderson: *(Front) Chisholm resident pulls over to let a horse and buggy pass on the road. [CWI]*

CNR replaced the original little station with a new larger station that included a waiting room and freight handling area.

The boarding house was built behind the station. Many workers passed through Fossmill to the bush camps and many single workers, like Stanley Anderson, worked for the company in Fossmill and needed a place to stay. The boarding house was a large building with steam heat and electric lights. Most of the workers' rooms were on the second floor but Peggy McGibbon remembers that "There were also some rooms at the back, downstairs, which were the favorite rooms." The boarding house also had a "big, big dining room, and at the east end there was an apartment with four bedrooms and a living-and-dining room for the cook and his family."

The boarding house residents were generally labourers who went home when there was no work, when they were homesick, or when needed by their wives or parents.

Peggy remembers that as a young girl she was somewhat intimidated by these seasonal workers. "Sometimes, when I was going to see my friend, the cook's daughter, I had to walk past a dozen men sitting on the steps. It was like running the gauntlet." Margaret Gleason says, "When we were in the school yard at night my dad kept an eye on us, and we had to be home by eight o'clock. There were all kinds of men in the boarding house nearby and he had four daughters, so he was always aware of what was going on."

It must have been lonely for many of these men who were away from their families. Myrtle MacDonald's daughter, Myrtle, recounted how after supper the men would sit outside the boarding house. "My mom would be singing while she rocked her children on the verandah—the men used to sit outside and listen to her sing. They said she was a beautiful singer."

Social events were often held at the boarding house. Rachel Denis remembered "dances, picnics and lots of anniversaries at the bunkhouse. Amedee used to play the violin. We would square dance and round dance."

Company store and office

West of the boarding house was the combined office, store, and private hotel. The office handled all of the cutting, milling, and sales records as well as the purchasing and local financial dealings of the company. The company head office was in Montreal. Custom orders were taken there and were forwarded to Fossmill.

The company store provided wholesale service to the jobbers in the bush, purchasing for the mill, and was a retail store for village and nearby residents. All in all, a considerable amount of goods were handled by the store, and sales from the store amounted to 14 % of the company's income. The store manager was Tom Robertson. He was helped by Morris Donahue's son, Joe, Stanley Anderson and others.

"I'll tell you a story about the store." Doreen Smith fondly remembers one Christmas. "We didn't have much money. I think I had 54¢ and I wanted to buy a present for my mother. I went to the Fossmill store but everything was marked above what I had. Then I saw—oh! it was very pretty—a little thing for mustard with a spoon in it, painted with a nice picture on it. Tom Robertson, seeing my predicament, said, 'Well now, what you have there will pay for that.' So I proudly came home with that as a gift for my mother."

Ivy Hall (Grant) remembers that her mother used to send her to Fossmill from their farm on the sixth concession, just north of Fossmill, for her favorite cheese and the mail. She also recalls that her father "cut out the bush and put logs across where it was swampy so he could take his horses and wagon for supplies at Fossmill." Local farmers sold food to the store, as well as directly to the

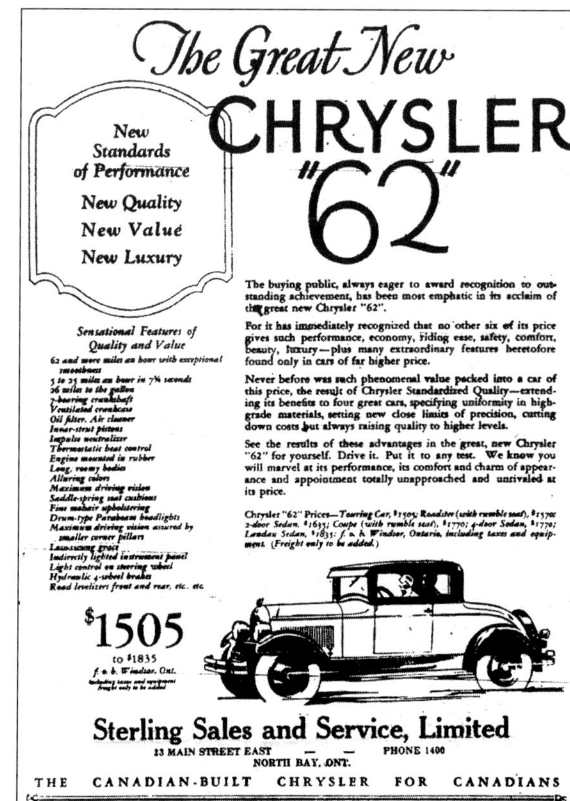

Company Store: *Villagers gather outside the Fassett company store. [LN]*

community, and also purchased items like tobacco from the company store, even though it was thought to be more expensive than the independent stores in Alderdale and Powassan.

Doreen Smith recalls that her father, A. J. Smith, used to "sell about 400 bags of potatoes to Fossmill each year in the fall. I sold some too, at 25¢ a bag—and the bag had to be full!"

Some Fossmill residents also sold goods to the store and boarding house. John Gleason's family sold milk. "They wanted a source of milk in the village. There was no milk close by so we bought several cows from the Cossar's and built a little barn at the back. My dog, Teddy, and I used to have to get up in the morning and get those cows. We would go down about a mile and bring them back up here. Dad would milk them before he went to work. The milk was for everybody. We used to sell most of it to the boarding house. In the morning we would have to strain all that milk out and my mother and I would take it around town. We had a little wagon, with a rack on it, to get to the far end of town. My dad eventually got a 1927 Chevy and later he bought a touring car and we used to put the cans in there and take them around."

"I washed the bottles and the separator every day," remembers John's sister, Nora. "We got about 8¢ a quart for the milk, and some people would come and want a pint. Well I would wash a bottle and fill it and give it to them for 4¢." Her sister Margaret says, "Mother would skim the cream off and sell that too. The milk that was left she would sour and make cottage cheese."

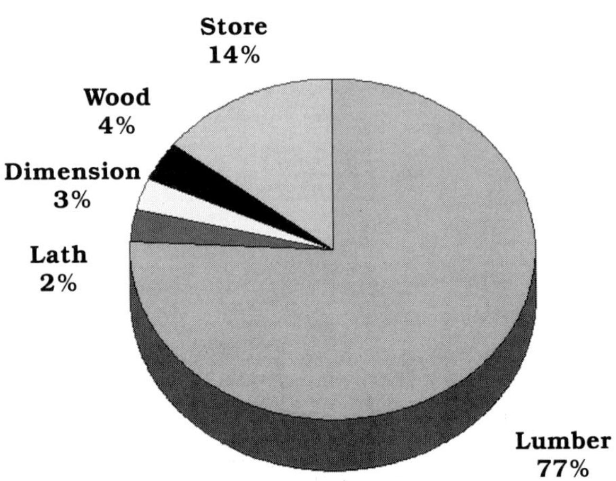

Company Income

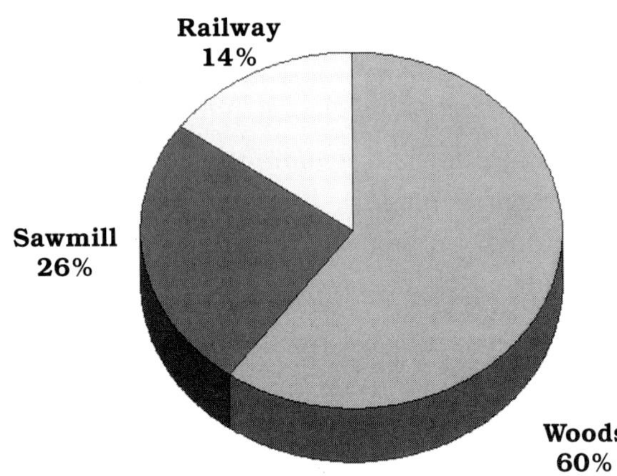

Company Expenses

Other services

Many services were provided by villagers from their homes. Albert Denis, the local carpenter, made toys for children, and made furniture and did house repairs for the community. One man made beautiful caskets. There was also a part-time shoe repairman and a part-time barber.

The closest doctor was in Powassan. Midwives delivered many of the village's children. One midwife was Gulia Huneault, Lawrence and Donat's mother. One resident remembers that she was a "big woman who usually wore a big white apron. She helped everybody with their babies."

Earl Orr recalls that one time the school was used as a hospital. The Health Department "set up an operating room right in the school and a team came down from North Bay. They took my tonsils and adenoids out. They took the tonsils and adenoids out of almost every kid in Fossmill."

It did not take long for Claire and Octave Denis to explore the village when they arrived in 1927, but it would take the French-speaking children time to become part of the growing community. "When we got there somebody introduced me," Claire says. "I had to ask my father what you said when you're introduced. He said, 'glad to meet you' or 'pleased to meet you.'"

Claire and Octave started school with teacher Ted Priest in September 1927. Claire says, "When September rolled around we had mastered only a few words of English, but then we had to do arithmetic and reading—it's complete immersion when you don't know how to say 'yes' or 'no.'"

Modern mill and yard of the Fassett Lumber Corp., Limited at Fossmill, Ontario

An Unbeatable Combination!

Logs from the famous Algonquin district---
Modern mill facilities equal to the best---
A fast and experienced service to wholesalers.

These are the basic reasons for the outstanding quality of Fassett Lumber and the dependability of Fassett' service.

From the time that the timber is cut in our limits until it reaches your yard, Fassett lumber presents a story of efficient workmanship and untiring efforts to provide an unequalled service in the manufacture of quality lumber.

The history of the Fassett Lumber Corporation Limited—its development, resources and methods is fully described in this issue. We simply add our desire to co-operate with our customers in every possible way toward the successful and profitable merchandising of our products. Your inquiries are invited.

We take this opportunity of extending to all our friends Hearty Christmas Greetings with all good wishes for the New Year.

FASSETT LUMBER CORP., LIMITED

Head Office
704 Bank of Nova Scotia Bldg., 437 St. James St.
Montreal, Que.

Bandmills and Planing Mills at Fossmill, Ontario

Fossmill school: *Girls skipping by themselves. They wore dresses winter and summer. Boys playing baseball in the yard to the right (also see photo page 10.).* [JD]

7

THE PEOPLE

"...like one great big happy family."

On the first day of school in September 1927, the teacher, Ted Priest, entered the school and surveyed the class. He had attended the Hambourg Conservatory of Music in Toronto when he was 17, before going to Normal School. After graduation he was lured to Fossmill, as their second teacher, by a $900-a-year salary. He was 21 years old, and this was his second year teaching in the backwoods of Ontario.

Standing in front of the class, Ted would have noticed Claire and Octave Denis and other new students from Quebec, and thought about the extra work needed to teach them English and assimilate them into the English school system.

First Communion: *Rejean Huneault died of kidney problems shortly after the photograph was taken. [LN]*

Girls: *Teacher Ted Priest, and the girls in his Fossmill class. [LD]*

Ted probably straightened his body and inflated his chest. He had learned early on that it paid to be imposing in front of 50 students, some of whom didn't want to be there. "He used to come in, in the morning, take his suit coat off, and roll up his sleeves," says Roger Bergeron. "He'd take the strap out of the drawer in the desk—pow-pow! He'd pound the desk two or three times with the strap and that was it for the rest of the day."

Ted was a tall imposing man who did not relish the role of disciplinarian—what he truly enjoyed was making music. He taught private music lessons after school and was invited to play his violin at both English and French gatherings. In school he introduced music to students who seemed more interested in games and sports.

"I could talk to you about Mr. Priest for a couple of hours," says Peggy McGibbon. "His story at Fossmill is unique. He was a big, tall, good-looking man, a wonderful musician, and a beautiful singer. What he did with those kids is something!"

She did recall, however, that Mr. Priest had to be tough on occasion. "He lined up as many as 20 students and strapped them one after another. When he came he was warned, well-warned, about the situation—but he was not a tough kind of person, he was gentle, kind and loving. I know, because he taught me music, but there was something that had to be done. Mind you, the strapping didn't go on for very long."

Ted discovered early on what the previous teacher had discovered. Many of the young people in Fossmill had formed into a group or gang where the older and stronger ones protected the younger and weaker from the authority of the adults and especially, to Ted's dismay, the teacher.

A curious aspect of this gang was that the apparent leader of this largely Catholic group was a Protestant—the son of an Orangeman—12-year-old Boyd MacDonald. Boyd was always in the forefront when the group challenged the authority of the teacher. One of the group recalls. "When I first started school Boyd wouldn't let the teacher touch me—there was no way!—He always helped people out." One of Boyd's sisters said Boyd was always "trying to help other people and doing it the wrong way." Sadly, Ted realized that to do his job he would have to deal with Boyd.

Ted Priest and Boyd MacDonald

"There was the ringleader of the guys that were giving the first teacher trouble," says Roger Bergeron. "Ted Priest told him, 'I'm the teacher here.' So there was a challenge between the two of them. Priest took off his glasses, rolled up his sleeves and said, 'we're going to settle this right now. You're going to come outside and we're going to settle this.' They had a fist fight and Ted Priest gave the guy a couple of good blows. That was the end of it after that—I'm telling you, a fist fight."

"Priest had to be tough because the kids

were tough," recalls Louis Demers. "I remember the fight. Priest grabbed him and put him to the wall." Peggy McGibbon was also there. "I was in the school when all this happened. I can remember the dramatic way Mr. Priest took his coat off. He was a big man about six-foot-two, good-looking, you know, and had some experience as a boxer—but he was not a boxer type. He was a musician, the son of a Baptist minister."

"I have a better story than that," Peggy recalls. "Mr. Priest told one boy to leave and he wouldn't leave. He said, 'well you have to leave.' The boy said, 'Well, you have to make me,' and so Priest did. We had those double seats that were screwed into the floor. The student said, 'You will have to drag me out of here.' He was about five-foot-ten or eleven and just sat there. Ted Priest picked him and the desk up and carried them out the door."

Of course disciplining students was not his only job; he was there to teach. "If they were going to go to school they were going to follow the rules and they were going to learn," says Peggy. "There were at least 45 to 50 kids in the school; it was wicked. But you know how he controlled it in the end?—by playing the violin and teaching us all to sing! There is nothing like music to calm the soul."

Apparently the music didn't have the desired effect on all of the students. Louis Demers says, "He used to play the violin, he was playing that classic stuff—we hated that stuff. When we wanted a break from school we would get after him, 'Bring your violin, bring your violin.' Oh boy, there was nothing he liked better."

Boys: *Teacher Ted Priest, and the boys in his Fossmill class.* [LD]

Ted Priest made a lasting impression on many students. The Denis family biography described Claire's early days in school and Ted Priest "the kind, patient teacher from whom she learned English." But he did get on the wrong side of some. One student says he "knew it all, nobody knew anything, no matter what, he knew more than you did." Another student said, "I was scared of him. When he came I was only a little thing. The first time I saw him he had this high hat and a cane swinging around. He was tall and I took one look at him and I thought 'holy gee', and I ran home crying, geez I hated him. I was sure glad to see the backside of him."

After his two years in Fossmill Ted Priest became a highly respected public school music teacher in Stratford, Ontario (see sidebar over). When contacted at age 90 he had fond memories of his teaching experience in Fossmill 70 years before.

Fred Hall and Fred Smythe

Most teachers stayed for just a year or two. Margaret Gleason was taught by most of them. "Fred Hall was nice; he was good with us kids; he would come out and play ball and other games with us. Mr. Smythe was a stern guy. He would let you get away with things for a while and then all of a sudden he would say, 'we're gonna catch you and we're gonna have a thrashing,' and on Friday he would line

Ted Priest

[AP]

When Ted Priest died in December 1997 at the age of 91 the local Stratford newspaper *The Beacon Herald* ran a story about him that said "he was bright, cheerful and always positive." It said that he taught music to all elementary and secondary school students in Stratford from 1930 to 1945 and all elementary students until 1967. Ted's daughter, Ann, was quoted. She said, "he was kind and gentle and very patient." The article indicated that his first teaching job was in 1927 "at a lumberjack camp near North Bay."

Daughter, Ann Pike, sent her dad's photograph of the Fossmill school (page 53) and the *Beacon Herald* photo of him above working on his prize wining wood carving, for which he was well known. Ted also wrote several music books for students. He made and played "Freddie the Fiddle" and it was almost as well known to students as Ted himself. David Priest, Ted's son, is a well known jazz musician in British Columbia carrying on the family tradition.

you all up there and give you a thrashing."

"Mr. Smythe used the strap once too often," one student recalls. "We ganged up on him. All week he would list the kids and then on Friday he would start giving the strap. There were four of us that day. Whoever went first pulled his hand back, and the teacher hit himself on his leg. The rest of us ran up and got behind him and put him down. We gave him quite a beating. He never told anybody and he never bothered us again. That was well into the year—he was using us pretty rough!"

"He gave me the strap," admitted Claire Denis (Grenier), remembering the time Fred Smythe caught her and her friends smoking. "The windows were on the south side of the school and he could see the outhouses—those diamonds they cut in the doors—the smoke went whoosh! It was Jeanne Maloche, Beatrice MacDonald and myself. He gave us each nine straps. I had great big welts on my wrist. Sitting at the table that night, my father sees the marks and says, 'What have you got there?' So I had to tell him." Claire's family biography notes that when her father discovered the reason for her swollen hands she "suffered further physical punishment."

Although many memories were of disciplinary actions, some students did recall the learning experience. Earl Orr, who went on to become a professional engineer, recalls that Fred Smythe "was bloody good. He controlled the class too. I remember that I was spelling after the first year, and I learned my times tables perfectly. He would be walking around the room and he would be talking to someone else. Then he would turn to you, 'six eights,' and you would fire right back 'forty-eight.' When he got you to do your times tables like that, then he didn't bother you anymore, and I learned that very early in the game. As a result, math came easy."

Joe Demeza

Joe Demeza, the teacher who followed Fred Smythe, was greatly admired by former

Fred Hall: *Third Fossmill teacher 1929-30.*
[CWI]

Fred Smythe: *Fourth Fossmill teacher 1931-32*
[CWI]

students. "He was an unbelievable teacher," says Roger Bergeron. "Kids were crying when he left. He would get us together at night and have a campfire and songs."

"Mr. Demeza was a good teacher," says Margaret Gleason. "That school didn't have just two grades to a room. We had one room with all the grades. We had junior primer and senior primer, junior first, senior first, and right up to senior fourth. By the time you got through senior fourth you pretty well had two years of high school. Well, I took my two years of high school with Mr. Demeza."

However, Peggy McGibbon recalls that he wasn't able to find much time for the senior students. "The first day of school he said, 'This is the course of study and here are the books,' and I sat in the back seat and did it."

Joe Demeza also had to discipline students on occasion. "One day Nelson Trudell got into a temper tantrum and was going to get the strap," recalls Lionel Martin. "Nelson said 'no!', and ran home to his house behind the school. Mr. Demeza went into the house, no one was home, and he hauled him out from underneath the bed and brought him right back to the school."

In a 1995 interview, Joe Demeza had fond memories of his work in the evenings supervising games and teaching music. He was paid $700 a year to teach 57 to 60 students. After teaching at Fossmill he became a principal, school superintendent, and eventually the Superintendent of the Ontario School for the Deaf in Belleville. (see sidebar over.)

Recess: *Fifth teacher Joe Demeza and some of his 60 students during recess at the Fossmill school.* [JD]

Class photograph: *Girls and boys of Joe Demeza's Fossmill class in the early 1930s.* [JD]

Joseph G. Demeza

Joe Demeza graduated from High School and Normal School in North Bay where on both occasions he was valedictorian. In 1933 at the age of 19 he acquired his first teaching position at Fossmill

Following Fossmill, he became principal of West Ferris Public School and earned 2 university degrees and his school inspector's qualifications. From 1947 to 1950 he was the first Superintendent of Public schools in Kirkland Lake. In 1950 he had the opportunity to work with deaf children at their school in Belleville where he became Superintendent.

His innovative and progressive work earned him an honourary Doctor of Letters degree for meritorious service to the deaf. He has always been active in music, church and school affairs in Belleville where he continues this work since his retirement in 1979. He was awarded a Canada 125 medal on Canada's 125th Birthday recognizing his outstanding community service.

The Fossmill villagers found time for a variety of leisure activities outside of the routine of work and school.

Fishing

As Fossmill was surrounded by many streams and lakes, fishing was a favourite pastime for many. Earl Orr says one of his best memories was fishing with his father. "I used to go up into the Park with my dad on the train. At the end of the day they would pick us up. I used to fish on all those creeks and streams. My uncle, Tom Cossar, used to take me fishing on Depot Creek. There was good fishing there and on the Wasi River. Sometimes we used to go to Wasi Lake for pickerel with the people from the church."

Proud: *Fossmill children show off their catch. "The girls were more tomboy than anything else," recalls Roger Bergeron. [MG]*

Roger Bergeron says, "On the weekend my dad talked to the fire ranger, Billy Tennant, about using his **speeder** to go into the Park. Bill said, 'Here's the key,' and we would head out. We stopped at Little Moose Lake where my dad had a wooden boat and oars. There weren't many fish there so we rowed to Big Moose. We caught nice four or five pound lake trout. My dad used to row when we were going upwind and I would troll at the back of the boat. When the wind was at my back my

Louis Demers: *Riding a speeder on the Fossmill Logging Railway. [LD]*

dad would fish and I would row. I didn't have to row hard because of the wind."

Swimming

Many villagers recall swimming in the Wasi River on the western edge of Fossmill. One villager said. "We went swimming a lot, at a bend in the river. There was a hole there, deeper than the rest. The boys and girls, having a hell of a good time."

Margaret Gleason remembers that Stella McGibbon "would be the leader and supervise us. We couldn't go unless she did. She would gather us all up and take us down there, sometimes twice a day. There was one deep hole and there was a shallow place where the little kids would paddle. That's how we learned to swim. We didn't have bathing suits; we used our underclothes. We weren't long learning to swim. She would be there and she would look after us and teach us."

On the tramway: *Children look at their father's handiwork. Generally they were not allowed on the tramways.* [DS]

Swimming hole: *Villagers take a dip in the Wasi River.* [CG]

Tobogganing: *The Wasi valley at Fossmill was ideal.* [RD]

Baseball

Baseball was popular in the 1920s. There was a ball diamond in the school yard for students, and in the 1920s the company sponsored a team that played teams from Powassan, Bonfield, Astorville and other communities. The company built a ball diamond about a mile west of the village opposite a CNR gravel pit. Roger Bergeron says, "The older guys, like Donat, had a baseball team with uniforms. Lawrence Huneault played behind the plate. There was a couple of guys that were good pitchers. One was a left-handed guy named Lalonde. Another was my brother Alderic."

Tennis and summer mischief

"We played tennis a lot," says John Gleason. "Down behind the Howard and the McGibbon houses they had a big tennis court. They let everybody play. We used to play there all the time. We used to get our rackets and balls from Mrs. McGibbon."

John Gleason also recalls going into the lumberyard to play. "The tramways were 20 feet high and we used to climb up there. They didn't want us to, but we'd do it anyway." Peggy McGibbon knew the company did not approve, but she says, "We would sneak up there and steal a ride on one of the tramway cars. Oh boy, that was a no-no! We had to do that on the sly. We weren't allowed to run

Skating lessons: *Raymond Donahue teaches Claire Denis to skate on the Wasi River. The steaming mill hot pond is in the background.* [CG]

among those lumber piles, mostly because they were afraid that somebody would light a match."

Winter

Bernice Martin preferred the winter months for playing with her friends. "In the summer the blackflies were really bad. We spent the summers picking berries, walking miles and miles and miles with our pails and our lunch. We'd be gone all day and come home with a pail full of berries. To me it was kind of boring. We couldn't wait for winter—we never felt the cold. Roger Bergeron's

Fossmill courting: *Amedee Laferriere and Rachel Denis. [GL]*

parents had a steel toboggan and it was never home. You could get on it and be down that hill in a split second. They always used to loan it to us. We all had sleighs but they wouldn't travel that fast. We used to spend hours out in the snow." Claire Denis recalls that Mr. Bergeron "used to take the used bandsaw blades from the mill and make toboggans."

Another man says, "In the winter we used to go skating and play shinny. There was no organized sport. We used to go tobogganing and skiing all the time." Peggy McGibbon says, "We kids were always glad when the mill stopped running. The pond would freeze and all of the young mill workers became temporarily unemployed, so you would have a number of teenagers with nothing to do but make beautiful rinks. We also had a lovely ski hill with jumps and everything."

Margaret Gleason remembered skating too. "The boys cleared off the creek, a big patch, and we'd have a big bonfire and we'd skate. We made our own fun."

Growing up

As young people grew up, leisure interests turned to parties, dancing, and dating. Many gatherings were spontaneous. Margaret Gleason says, "In the winter we always had card parties, every weekend, from one house to another. Sometimes the young people would just get together for a dance. We'd have a house party. We'd just say, 'Dad can we have a party tonight?' and he'd say, 'yea', so we'd just roll up our Axminster rug and move the furniture—Dad would call the dances, and that."

Her brother, John, says that they would socialize all over the area. "We used to be back and forth to Chisholm all the time. We used to go to parties mostly to Chiswick and out to dances at the Orange Hall. Of course there were all the guys and the girls down around Wasing and that. Out there, they'd make a party out of anything."

Nora Gleason remembers that they also had dances at the school. "Lawrence Huneault was a caller for the dances. The least excuse they would have a dance. They had a violin player. Everybody made sandwiches and stuff.

Well dressed: *Fossmill's Elva MacDonald in the latest fashions. [MC]*

We made our own good times."

"We had square dances and round dances at the bunkhouse and lots of anniversaries," said Stella McGibbon. "The dances were the thing. Dances all around the township—big dances at the boarding house, especially when the logs were out in the spring."

Doreen Smith recalls that there was "mostly square dancing. There was a little round dancing, the odd waltz, fox trot or two step. I never did that. I went for the square dancing."

There were several fiddle players and dance callers, but the singing idol of many was the handsome young Alderic Bergeron. His brother, Roger, recalls how "he used to sing and play the guitar, with his mouth organ strapped around his neck, any time there was a concert." Bernice Martin recalls that "Alderic used to play the guitar and sing. He was our idol, imitating Wilf Carter."

"I met Alderic at a dance in the boarding house," reminisced Doreen Smith who went out with Alderic. "He would walk to my house from Fossmill or I'd go and meet him and we'd come back here and play cards with my parents and we'd go to the odd dance together." Other young women were attracted to him as well. Peggy McGibbon remembers that she went out with Alderic for a while.

City visitors: *Claire Denis (centre) with her sisters Juliette and Laurette when they were visiting the family in Fossmill.* *[CG]*

"Culture"

"Mrs. Howard was a good singer and so were Ted Priest and Alton Douglas." Peggy McGibbon recalls that, besides popular music and dances, some people were also interested in classical music, opera and theatre. She remembers, "There were people in Powassan, who were singers. They had a group that put on the most beautiful, excellent musicals. They went all over with them. They always came to Fossmill. The shows were like an operetta. They did Gilbert and Sullivan, *The Gypsy Rover*, that kind of thing."

Movies and radio

Movies and radio, the expanding entertainment media of the 1920s, also had an impact on Fossmill. The Reverend Charles Reid, an Anglican minister from Mattawa, came to Fossmill to show movies. "I went to all of them. Most of them were Charlie Chaplin movies," says Margaret Gleason. "We thought it was just wonderful whenever he came. All dead silent—Charlie Chaplin was funny."

Roger Bergeron provided another perspective."They were silent pictures—free admission, but he had a silver collection." "Five minutes before the end of the movie, he would stop the movie and make the collection. He wanted the nickels and dimes. A lot of them were repeats, but everybody would go anyway—silence, utter silence."

Batteries were necessary for the incredible talking box, the radio, that brought music, news, stories, and education to Fossmill. "I remember our old battery radio, a Philco shaped like a big arch," recalls Earl Orr. Peggy McGibbon recalls listening to "plays and Amos and Andy, and Jack Benny in his early days. The best stations in the early twenties were from New York, Pittsburgh, and Schenectady, New York."

Reading

Reading was another pastime, one villager remembers, "My mom read a lot to us. She got the books from my grandmother. She would also read the Bible to us all the time. That was our bedtime story."

"I read the Toronto Star. It had a 16-page pull-out section each week from an ongoing novel," says Nora Gleason. She remembers how important reading was to her and how she "used to live for that novel, and the comics, of course."

Women's place

In the 1920s, the place of women in society was going through many changes. Since World War One, when women had taken many of the jobs traditionally reserved for men, women had been fighting for and achieving many rights. By 1922 women could vote federally, and in all the provinces except Quebec. In 1921 Agnes Macphail became the first woman elected to parliament, and in 1927 women were recognized as "persons" in Canadian law. The impact of these and related changes were slow to filter down to Fossmill.

A few women worked in the boarding house or as paid housekeepers. However, most women stayed at home and worked hard rearing children, cooking, washing, and gardening. One villager recalls, "My mother never wanted to go out, not until the day she died. My dad used to go everywhere." Some women, like Stella McGibbon, learned how to drive, or like Margaret Howard, became active in the United Church Woman's Auxiliary.

Young people in Fossmill, however, were influenced by the changes taking place in society. Rachel Denis' sister, Juliette, from Ottawa, caused quite a stir "on her visits to Fossmill, wearing fashionable short skirts and hairstyles that were prevalent in the roaring 20s," says the family biography. "She drew admiring glances from the young men at the mill and indignant stares from local churchgoing women. Her father was concerned for the good family reputation."

Claire Denis was punished for experimenting with tobacco and it can be assumed that her friends received similar treatment. Claire looked beyond Fossmill and wrote her sister who got her a job as a housekeeper in Ottawa. She left Fossmill when she was 15 years old (see profile page 211).

Prohibition

Although prohibition was still in effect in Ontario until 1927, drinking alcohol was still a regular pastime. "The people at the bottom of the hill made moonshine. You could smell it in the winter," says Peggy McGibbon. Another villager mentioned how his father bought moonshine regularly and another recalls that "One man couldn't handle his liquor and would get out of control. He tried to beat up his wife and she ran from house to house with him after her, and people hid her. When he was sober he was the nicest person around. Of course, he had to have a few drinks with the boys on payday. They shouldn't have given it to him. It was sheer luck that he never managed to kill that poor woman."

Ted Priest, the school teacher, recalled playing at a party in the late 20s, at which his accompanist was so drunk he could hardly play the piano.

In opposition to this drinking, many women were active temperance supporters. Earl Orr recalls his mother's strong stand. "My mother was involved in the temperance movement. I remember she got mad when the women used Labatt's ashtrays to hold pins at the quilting bees!"

Switched: *Young man and woman experiment with changing roles.* [LN]

Back porch: *"Aunt Maggie" Donahue (right) visits with a neighbour. Maggie was a well loved elder in the community. Myrtle MacDonald said, "The first flower that grew in the spring was picked for Aunt Maggie. It didn't matter if it was a strawberry blossom or a pink cherry blossom, we'd take it to her. She was a special lady."* [LN]

Accidents and sickness

Accidents, sickness, and death were common events in the village. "I remember a few babies dying in Fossmill. My dad was handy as a carpenter and he made caskets for the people there," says Roger Bergeron. He also recalled that one "young lad" had an epileptic seizure and drowned in the Wasi River.

Another villager recalls that "Rejean Huneault died of kidney trouble. He was just a little fellow." (see photo page 115) Margaret Gleason remembers the deaths. "At that time you didn't get to a doctor very often. There was a lot of little babies died and of course there wasn't any funeral parlours so they'd make their own little coffins and line them. Us school girls went to see every one of them." She also says that one time "an older kid got up and got a bottle of pills and fed it to the baby while the mother was out hanging up clothes—the baby died."

Crime

Fossmill was also touched by its share of petty crime. "My grandfather bought his first car, and somebody stole it from his garage," recalls one villager. "They ran out of gas and the cops picked them up." Ironically, the thief later married one of his granddaughters.

After the confrontation with Ted Priest, Boyd MacDonald went on to become a legendary figure in Fossmill. His defiance of authority and general mischief led to appearances in front of judges and time spent in jail.

"He wasn't bad—like a ten year old. You see something you take it. To him—it's there, well let's use it—that was his thinking," recalls one resident. "The very first time he went to court, my dad drove him and his mother. They went up in front of the judge and he questioned them. I think they had him evaluated and that was it."

"He would do things, not for himself. He'd do things for others," recalls Louis Demers. "He'd go and steal a bunch of turkeys around Christmas time and he'd give one to everybody—nothing for him." Another villager said, "Any trouble he got into, he owned up to it, and if there was three or four in the bunch, he took the blame for everybody."

Boyd was befriended by a number of people who had a great affection for him. One local resident remembered. "My dad did something for Boyd one time when he was in trouble and from that day on he was friends with my dad. He was quite a character. He was a funny guy. He took a notion to my dad and he would do anything for him."

One time, Boyd was working for the Ontario Northland Railway when a shipment of prize horses from out West was stopped overnight on the way to the Canadian National Exhibition. Apparently a local farmer had a team of old horses, and Boyd exchanged them for the prize ones. The farmer said he could "hardly hold them." The police did an investigation into the exchange of the horses but they never discovered who had switched them.

"My mother had a soft spot for Boyd," says Margaret Gleason. "She always said that Boyd got blamed for a lot of things he never did—and he did get blamed! I guess, because he was a bad little devil, people believed the other people. It wouldn't matter if he said he didn't do it or not. He got harsh treatment which doesn't help one little bit."

Margaret also told another touching Boyd story. "This big area school fair had prizes for different things, including maple cream. Boyd made some and it turned out hard but he put it in a paper bag and took it to the fair and they gave him first prize. My sister laughed about that so many times, about him getting the prize when all the others had nice plates with their candies all neatly displayed."

Conflict and challenges

Ron Brown in his book *Ghost Towns of Ontario* writes that Fossmill "in many respects suffered the cultural division found in many lumber communities in Ontario and Quebec. While mill and bush workers were French, staff supervisors and executives were English. The separation was physical as well as cultural."

On the surface, Fossmill contained all the ingredients of the conflicts evident elsewhere in Canada in the 1920s: French-Catholic versus English-Protestant, rural versus town, and managers versus workers. The communities of Chisholm and Fossmill responded to these challenges and divisions in their own unique ways.

Religion and language

One source of tension in Fossmill was the relationship between Protestants and the Catholics and between the French and the

Wasing United Church: *A gathering in 1929 includes many Fossmill and Chisholm residents mentioned in this story (see opposite).* [CWI]

English (some Catholics were also Irish and Polish). *Our Chisholm Story* notes that in the early days when Catholic Mass was held once a week in the boarding house, Father Gravelle, the Parish Priest, "welcomed everyone regardless of their faith." Most Protestants, however, went to the Wasing United Church.

"Everybody was friendly until they built the Catholic church," recalls one Protestant villager who had many friends among the Catholic children. "Everybody went to the same school, but as soon as they built the Catholic church, I don't know, there was a difference. They had to get to bed at night to get to Mass in the morning and this sort of stuff, there was a big difference."

To overcome this difference, Protestant children joined their Catholic friends for evening prayer sessions. Protestant Margaret Gleason says, "I went to the Catholic Church all the time and I got so I knew the prayers as well as the other girls did. My dad was pretty strict and you had to be in by eight o'clock but if you went to prayers at least you'd have a half an hour or so more to be with your friends." Margaret's parents and other Protestant parents did not object to this.

Stella McGibbon remembered looking for her children one evening. "I went to Allan Gleason's house because they had two little girls the same age as mine. Mrs. Gleason said, 'They were here, running around looking for something to put on your girls' heads. I bet they went down to the chapel.' I went down and stood around the chapel door till the service was over. Sure enough out come my kids with the Roman Catholics—just the same."

Father Gravelle, the Priest, was known for his non-sectarian attitude. Margaret Gleason remembers that he was "one of the nicest men I've ever known. He was just the same with us Protestants as he was with the Catholics. There was no one that my father liked any better than Father Gravelle—he didn't talk religion to us. If my dad wanted to find out anything about law, or something that he needed to know, he used to go down to see Father Gravelle."

Legend for the Wasing Church photograph opposite:
1. Tom Howard
2. Agnes McFaul
3. Mrs. Howard
4. Mrs. McFaul
5. Gordon MacDonald
6. Stella McGibbon
7. Alton Douglas
8. Lois and Peggy McGibbon
9. Doreen Smith and Irene Howard
10. Max Smith

Many nationalities were represented in Fossmill. The majority were French-Canadian, but Allan Gleason was American; the blacksmiths, Joseph Sliwka and Joseph Kozlowski, and Louis Polichuk were Polish; Morris Donahue was Irish-American; Joe Kunkel was German-Canadian and Myrtle MacDonald was Native. Of course the Anglo-Canadians such as the Howard, Summers, McGibbons, and Douglas families, were also members of the community.

As mentioned earlier, the French children had to learn English because until 1927, regulation 17 required all schools in Ontario to offer classes only in English. The English students helped the French students to learn English. One student recalled, "The older girls or boys would teach these little French kids their sounds, and you know, before the year was out they'd have gone through their ABCs. You'd be outside playing and your friends would be French. You'd say, 'Talk French to me,' so you could learn French but it was easier for them to speak English than to try and teach us a little French."

Margaret Gleason says, "I chummed with the French people all my life—Mother wouldn't have it any other way, because where she was from, down at Greece's Point in Quebec, she was raised among the French. Mother could switch to French as quickly as she could to English and so could my older brother Lester."

Jack McGibbon hired the men for the woods operation. His daughter Peggy recalls that he "delighted in the French people, learned to speak French, and hired mostly Frenchmen because they were the best lumberjacks. We never learned to speak French the way the French did. My father was bilingual. The French people said, 'You didn't know he wasn't French.' But he didn't learn French until he was a teenager because the French learned English, but the English didn't learn French. At Fossmill all of the French children learned good English, but we didn't learn much French, unless we made an effort. For example, I had a French boyfriend, so I said, 'Speak French, talk French,' so I could learn. I went to Maggie Donahue for two years. She taught me French twice a week. She came from Fassett and taught in a separate school there."

"He was quite a hockey player": *Willie Maloche (junior) became a professional hockey player and coach, but continued to work on the log pond at Kiosk during the summer.* [RM]

Two stories demonstrate the occasional tension felt in the village. Nora Gleason recalls one incident. "There was a hockey match on the pond and this Maloche—he was quite a hockey player. They had two teams, the English and the French teams, and he wanted to be on the English team. He said to the Donahue boy, 'Can't I be English for just one day?' Donahue replied, 'Hell no, you're still French and you're playing on the French team!'"

Earl Orr still has the scar to show for another incident. "I remember as a kid we used to have the odd squabble with French kids. We used to have stone fights. I got a scar right here from one. You'd have a stone fight and then you'd be out in the snow playing the next day. It was just—it seemed to be when you had problems you just divided. I had fights with some of the English guys too but there weren't too many English there."

Some older women who stayed home had trouble learning a second language. One villager said that when her mother died in 1973, "she could hardly speak a word of English because Fossmill was all French and then Kiosk was French too."

Lionel Martin said with a smile, "My grandmother wouldn't speak English. My great grandmother lived with them and she would not speak French—but they spoke to each other and answered each other—they were two stubborn women!"

Rural community and villagers

Another source of tension was the relationship between the people in the surrounding community and the mill workers and their families in Fossmill. Fossmill's development benefited the farmers in Chisholm Township in many ways. "I don't think you would find any family in the area whose parents or grandparents didn't work at Fossmill at some time," says Peggy McGibbon. "I drove those roads with my father. I knew them all so well because we would go and buy hay and oats and hire men. Fossmill contributed to the development of the township and area for years. Farmers sold produce, people got jobs, we got new roads. For example, there must have been 20 to 30 cars in Fossmill before there were hardly any in the township."

The workers in Fossmill were wage earners and although they did not make high wages by industrial standards, they had a steady income. In contrast the rural farmers who survived on marginal lands may have felt envious of the Fossmill workers who, among other things, could afford to buy cars.

One child of a local farmer, who was asked if there were any negative feelings towards the Fossmill workers, replied, "I couldn't really tell you that for sure—I know there was a feeling, more or less, that they didn't seem to have a respect for property and, of course, they all had a little more—as far as money goes—than the rest of us had." One villager commented, "Well the farmers sort of stuck to themselves. We were all right to sell stuff to, but that's about it."

Others interviewed denied any problem at

all and pointed out that many people from the surrounding community had full and part-time jobs at Fossmill. Some office staff lived in Powassan and stayed in Fossmill during the week and some workers lived in the township. Joe Kunkel, for example, lived down the road and worked at Fossmill for years and many of the teachers that worked at Fossmill from 1924 to 1947 were from the area. The Graffs and Smiths and others worked in the bush. Tom Robertson, the store manager, was from Chisholm, as were several of his helpers, including Stanley Anderson. Stanley sold large quantities of maple syrup to the village from his family farm. Jesse Battrick and A. J. Smith sold potatoes to Fossmill.

Bernie Cossar, who still lives nearby remembers, "My uncle worked in the stable at Fossmill but the farm was primary for him. My dad worked in the picket factory in the mill for a while." The Cossars also provided eggs and cream to Fossmill on a regular basis. Max Smith remembers that his dad "used to supply Fossmill with meat, sometimes through Ross Ellis, the local butcher, and he sold eggs and oats too."

Many members of local farming families married people from Fossmill. For example Tom Cossar, Bernie Cossar's father, married Areta Olimer, the daughter of the boarding house cook (see photo of Areta and Tom on page 150).

"Maybe the bosses made money," says Doreen Smith from Chisholm, "but most of the people—you know that song about, 'He owes his soul to the company store'?—that was the case. They all had big families. They all had to buy their stuff at that store. I know that when my Fossmill boyfriend, Alderic, took me to a dance at the boarding house he got his dad's car and enough gas to come down to get

Doreen Smith: *Waiting at the Wasing Station with cream from the family farm.* [DL]

Subsistence: *Many local farmers survived on marginal lands and the crops sold to Fossmill provided welcome cash.* [CWI]

me and go back—that was a big thing. The people in Fossmill weren't any better off than us—they weren't as well off—we had all our own stuff."

Managers and workers

Two other sources of tension were the relations between different levels of company management and between the management and the workers. The forestry students reported on the hierarchy of power in the Fassett Lumber Corporation and in Fossmill. "The organization is fairly simple and direct. The president of the Company, Mr. Charles Dale is the nominal head. The vice-president and managing director is Mr. S. J. Staniforth.... He spends most of his time at the Montreal office, but makes frequent visits to Fossmill to see the operations personally. Directly responsible to the managing director is Mr. T. B. Howard, assistant manager, and Mr. C. H. Skelton, Secretary-Treasurer. Mr. Howard is responsible at Fossmill. Under him comes the mill superintendent. The woods superintendent visits the camps and oversees the operations."

A diagram of the organization placed the woods superintendent under Tom Howard, but Peggy McGibbon disagrees. "Howard was not over the whole thing. It was completely divided. He looked after the mill operation and my father looked after the logging operation. My father's position was Superintendent of Forestry. They were both autonomous in their own right and never interfered with what the other did—Mr. Howard did, once!"

She says that they worked as "very independent men" and explained that this "was a great way to keep the men feeling proud of their work, and feeling that they were not under anybody. My father would not tolerate Mr. Howard saying anything to him about logging, and Mr. Howard was the same way—nobody could tell him about the mill and yard operation, and that was fine, everybody was happy with that. Every section of the mill had its own workers. They were independent and nobody said anything to them."

Managers and workers: *Workers trim the end of a log while Tom Howard poses for the camera.* *[APA]*

Peggy commented on the different levels of status. "In the lumberyard for example, there were the Bedards. They were all more or less related. They ran the yard, getting the lumber there, piling it and loading it." The men who worked on the railway were "another layer of workers." She said there was a "pride if you ran one of those machines."

About her father's relationship with the men, Peggy also noted that "It wasn't really a class thing either, it was that my father was in a position of authority and as such you had to maintain a certain distance—'familiarity breeds contempt' is the phrase. If you are in a position of authority over people, you cannot be too friendly. You have so many men working for you."

Apparently though, Jack had a close relationship with the lumberjacks and socialized with them. Louis Demers says that Jack was "the **bush walker**. He would pass around the camps. He would sit down with the guys and grab a big dish and fill it up with molasses and eat it with a chunk of bread—he loved molasses!"

Max Smith says Jack was "a quiet man" who concerned himself mostly with the bush. He recalls that Tom Howard "was more of a going concern. He was all over in through here. He used to travel to North Bay and Powassan by car a lot. He'd run off the road and we'd draw him back on all the time. He was a pretty nice guy, more in charge."

Germaine Demers, who worked for the Howard as a housekeeper recalls that Mr. Howard didn't take much time off from work. "With him it was his work. He spent a lot of time in the office."

Donald Staniforth says that "Tom was a doer, if you know what I mean. Say, in the mill, something would go wrong, he wouldn't delegate, he would go and do it himself. He wasn't afraid to work, I've seen him take out slabs. You'd see him working around with his hands."

Manager's families

Stella McGibbon and Margaret Howard were active leaders in the community and their children mixed freely. Margaret was active in the Wasing Church. She founded and chaired the first Women's Auxiliary. Stella taught Sunday school, was interested in sports and took an active role in the recreational activities of the Fossmill children.

"We lived up there on the hill, the Howards and us, and we didn't mingle, I might as well tell you we didn't mingle. It was the position our husbands had," Stella said. "My social life was mainly playing cards. My friends came to my place because I had children. They were the office staff, like Mr. Howard, Mr. Summers, and Mrs. McFaul, who lived above the office. That was my social life—it was good."

Peggy McGibbon recalls that she "was the boss's daughter. I was 'Jack McGibbon's daughter,' and there were things you didn't do; there were places you couldn't go. I couldn't go picking berries—the rest of them went flying along. We were to do things in a certain way. I don't think there was as much freedom. We were brought up very strict and you certainly remembered who you were and you acted in a certain way. You had to perform, you were expected to do well, you were expected to act well, you were expected to be something, and it stuck with me."

The average wage of a mill worker was around $900 per year in the 1920s. Woods Superintendent Jack McGibbon made around $1,800 per year, a wage still less than Canada's average 1929 income of $1,900. The company managers enjoyed indoor plumbing, electric lights, a tennis court, and better cars, but the differences in living conditions and life styles were not huge. One villager remarked, "Back then, nobody had anything much better than anybody else, not even what they call the high society." Peggy McGibbon reported that after she moved from Fossmill to Powassan her mother used to say, "It's very difficult to come from being a big frog in a little puddle to being a little frog in a big puddle."

Prosperity and optimism

The late 1920s was a time of relative prosperity and optimism. When villagers were asked if they were happy with their life at Fossmill, the answer was generally "yes." Villagers were aware that the men in the bush were away from home a lot. The mill workers had homes and families that they saw regularly. There was also a camaraderie on the job that was satisfying.

On one hand it was clear that working for the company was a way of life for the mill workers. Many did very little other than their six ten-hour shifts a week, at low wages, with only a couple of holidays a year. It was a stable way of life, with enough to live on. The men felt they had respect in their jobs which they did well and which provided status in the mill and the community.

On the other hand some said that the company was fortunate to have the employees they did. One worker said, "I think that everything people got from the company, they earned. The company had good loyal employees that would drag themselves to work in the morning even if they were sick—of course, if you missed a day, you missed a day's pay, that was it! Jobs came from father to son. It wasn't a bunch of people that would come in and leave."

By 1929, the people in Fossmill had forged their own unique community out of the woods of Northern Ontario. Margaret Gleason sums it up when she says that Fossmill "was like one great big happy family. If somebody was getting married, they were there to help you. If you had troubles, like somebody died, they were there to help you. They weren't sitting on your doorstep but they were there. It was just like it was their family that was in trouble. It was wonderful."

Life in Fossmill was good, but eight weeks before Black Tuesday, October 29, 1929, the day of the Great Stock Market Crash and the beginning of the Great Depression—things began to turn bad for the Fassett Lumber Corporation and Fossmill.

Railway workers: *(Left to right) Bill Gauthier and brakemen, "Man" Martin and Gordon MacDonald in front of the Barnhart loader.* [CWI]

8

Fire and Depression

Fire and Depression

"There was this glow in the sky all over."

On September 2, 1929, around 2 p.m. Jack Huneault was working in the loader, clearing the last logs from the previous year's cut from the log dump at the east end of Moose Lake. The loader crew was too absorbed in its work to notice that a spark had escaped from the loader's spark screen and had settled in the dry **slash** by the railway. By the time they noticed, a full-fledged fire was in progress. Jack signaled the locomotive engineer to pull the loader out of the area and they rushed to report the trouble to the Fire Ranger at Fossmill.

After crossing the bridge at the creek east of Little Moose Lake, near Camp One, they came upon Railway Superintendent Morris Donahue and his crew, who had discovered another fire.

Both fires were reported to Bill Tennant, the local fire ranger, around 3 p.m. The *North Bay Nugget* reported that the fire "ate a passage along the company railway spur leading into the woods, and in its progress destroyed a bridge. The danger became so great that the mill was closed down and all hands diverted to fire fighting."[14]

Bill Tennant called in four more rangers and 70 men were pulled from the mill to rush back to Little Moose Lake to fight the fires. By that time the flames had consumed the

bridge and burned down Camp One. A west wind fanned the flames and the fire spread quickly east into Wilkes Township and into South Chisholm. Most of the area had been logged over and the dry brush left behind was ideal kindling for the growing fire.

To add to the problem, two rangers had to be dispatched the next day to fight a 300 acre (122 h) fire that had started at the site of an old Fassett cut near the west end of Tea Lake, in Ballantyne Township.

As the original fires grew, it was clear that they would burn for days. There was some concern for the village and the mill. Roger Bergeron recalls that the CNR brought in a "special train with some coaches on it ready to evacuate the village." He remembers that the fire was so close that during the night villagers "could see this glow in the sky all over."

The fire burned for 11 days consuming 4,100 acres (1,660 h), including 65 acres (26 h) of uncut forest as well as the railway bridge, 1000 saw logs, two old camps, one new logging camp, and some lumber. Sparks from the company steam loader and locomotive were found responsible, and the Fassett Lumber Corporation had to pay the full cost of the fire-fighting. In spite of the setback, Fassett still operated one company camp and six jobbers' camps during the 1929-30 season.

The fire created problems for the logging camps beyond the burned bridge in the Park. The *North Bay Nugget* reported that the loss of the bridge "cut the camps off from the source of food supply and necessitated transshipping from **lorry**-to-lorry at the break in the line."

Spark screens

The forestry students had been overly optimistic the year before when they wrote that there had been "but one small fire [on the Fassett limits]....It was caused by sparks from the engine, but was only small and soon burnt itself out in an old burn....The Company does not take any measures to protect the limits from fire, beyond using **spark arrestors** on the engines and the stack of the mill."

Many logging companies were concerned with the methods used to prevent fires from their equipment because the spark screens, or arrestors, on locomotive smokestacks clogged up with ash and reduced the efficiency of the engine, leading to increased fuel costs. To solve this problem, many different smoke stack and screen designs had been used over the years.

Earlier in the century, the Cleveland-Sarnia Lumber Company operated a Shay locomotive on its limits near Diver, Ontario, north of Fossmill. A forestry student report on the company reported on its solution to the problem. It said that "No chances are taken with the little engine which literally threads its way among the trees. With the approach of warm weather the grates are removed [from the firebox] and oil burning apparatus

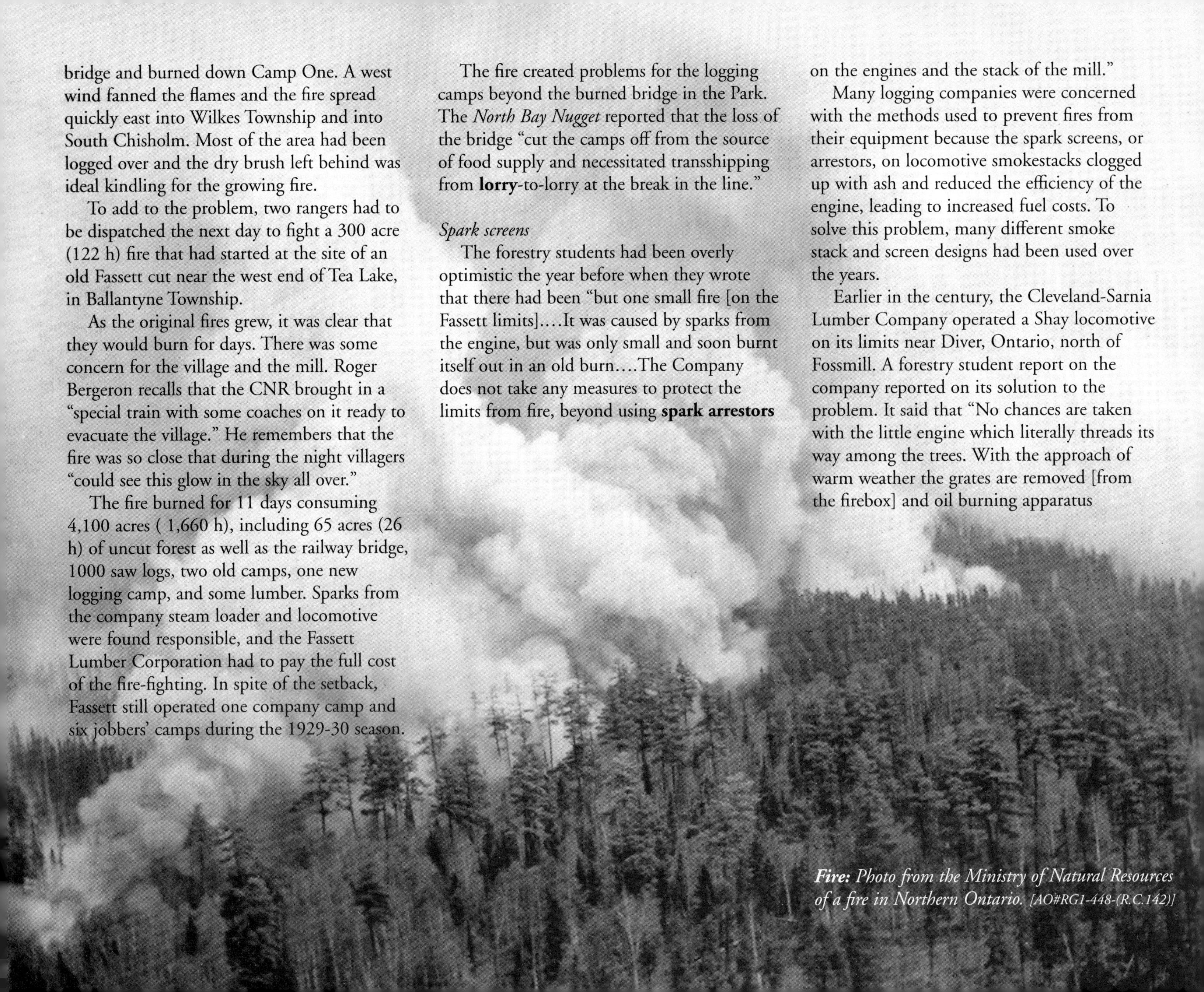

Fire: *Photo from the Ministry of Natural Resources of a fire in Northern Ontario.* *[AO#RG1-448-(R.C.142)]*

substituted. Thus there is no chance of fire from this engine, and it does not require attention from the fire rangers."[15] The Fassett Lumber Corporation chose to stick with spark arrestors, and the following year this would raise the ire of an improved Ontario Fire Service.

The Fire Marshal's Act

The problem of preventing and fighting forest fires led to the 1917 Fire Protection Act, and the development of the Ontario Fire Service in 1924. In 1930, the Fire Protection Act was replaced by the much stronger Fire Marshal's Act. It allowed for inspections, required fire protection devices to be installed on equipment, and allowed for the collection of the cost of any forest-fire-fighting from offending parties. Hundreds of rangers were assigned to oversee the protection of our forests. A fleet of airplanes was established and fire prevention and education were enhanced. The Ministry of Lands and Forests trained the fire rangers to prevent, fight, and report on fires.

The local ranger in the Fossmill area was Bill "Billy" Tennant of the well known Tennant family on nearby Wasing Road. His fire-rangers' cabin was on crown land just to the east of Fossmill. Billy Tennant's name appears on numerous fire reports in the provincial archives between 1924 and 1952.

In the spring of 1930, Billy Tennant, armed with the new powers granted him by the Fire Marshal's Act, was prepared to act quickly to prevent a repeat of the previous year's disaster. On May 6th, he extinguished a two-and-a-half acre fire along the railway in Wilkes Township, and instructed the company to stop burning debris along the railway. On May 9th, he extinguished a small fire at the mill site, and, becoming concerned, suspended the mill operation until an inspection was completed. It was concluded that the fire was caused by locomotive No. 51. The next day a spark from No. 54 started a fire along the railway south of Moose Lake in Ballantyne Township, burning 460 acres (186 h). It took four days to extinguish the fire. Billy acted quickly under his new powers and restricted the company to night work, presumably so that the crew could see sparks easier, and identify fires earlier.

"I used to get a chance to go on the train at night sometimes," recalls John Gleason. "My dad would have to go in for something and I used to go with him. I used to sit right up on the water tender. You really saw the fires, saw them jumping from tree to tree. I don't have any proof, but I imagine they were started by the engine. You know, they used to have screens on the stack, but the odd bit would get out."

Harsh words

In his 1930 Annual Report the new Algonquin Park superintendent, J. H. MacDonald, had some harsh words for the Fassett Lumber Corporation. He wrote that "There is, of course, the operator who is very loath to comply with any and all regulations, especially if there is any monetary outlay entailed. This man or outfit may be dealt with now under the existing legislation by closing down his operation until willingness is shown to live up to a reasonable standard of fire protection. A case in point might be cited: that of the Fassett Lumber Corporation who were our most serious offenders of the Fire Act during this season. They operate a private logging railroad on which there are many heavy grades and over which trains of as many as 16 cars are hauled by locomotives which have seen better days. The consequence of such a combination was that fires were occurring along the right of way. Two of these reached such proportions that, despite

Fire ranger: *On the lookout for trouble along the railway.* [AO# C7,17001]

Fire Reports: *Reports on the major fires on the Fassett Limits.* [AO]

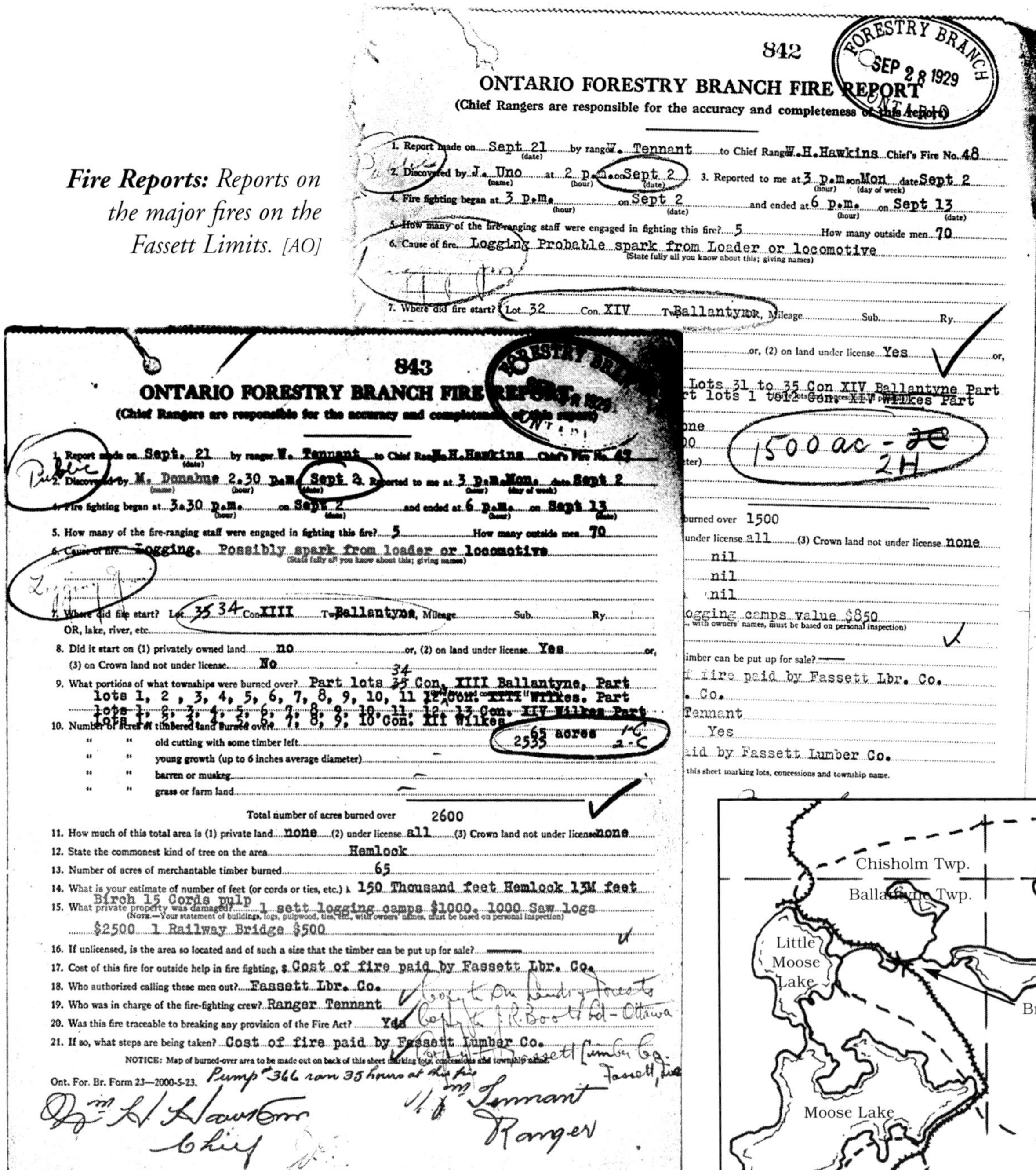

842

ONTARIO FORESTRY BRANCH FIRE REPORT

(Chief Rangers are responsible for the accuracy and completeness of this report)

FORESTRY BRANCH SEP 28 1929 ONTARIO

1. Report made on Sept 21 (date) by ranger W. Tennant to Chief Ranger W.H.Hawkins Chief's Fire No. 48
2. Discovered by J. Uno (name) at 2 p.m. (hour) on Sept 2 (date) 3. Reported to me at 3 p.m. (hour) on Mon (day of week) date Sept 2
4. Fire fighting began at 3 p.m. (hour) on Sept 2 (date) and ended at 6 p.m. (hour) on Sept 13 (date)
5. How many of the fire-ranging staff were engaged in fighting this fire? 5 How many outside men 70
6. Cause of fire Logging Probable spark from Loader or locomotive (State fully all you know about this; giving names)
7. Where did fire start? Lot 32 Con. XIV Twp. Ballantyne, Mileage Sub. Ry.

843

ONTARIO FORESTRY BRANCH FIRE REPORT

(Chief Rangers are responsible for the accuracy and completeness of this report)

1. Report made on Sept. 21 (date) by ranger W. Tennant to Chief Ranger W.H.Hawkins Chief's Fire No. 47
2. Discovered by M. Donahue (name) 2.30 p.m. (hour) Sept 2 (date) 3. Reported to me at 3 p.m. (hour) Mon. (day of week) date Sept 2
4. Fire fighting began at 3.30 p.m. (hour) on Sept 2 (date) and ended at 6 p.m. (hour) on Sept 13 (date)
5. How many of the fire-ranging staff were engaged in fighting this fire? 5 How many outside men 70
6. Cause of fire Logging. Possibly spark from loader or locomotive (State fully all you know about this; giving names)
7. Where did fire start? Lot ~~35~~ 34 Con. XIII Twp. Ballantyne, Mileage Sub. Ry. OR, lake, river, etc.
8. Did it start on (1) privately owned land no or, (2) on land under license Yes or, (3) on Crown land not under license No
9. What portions of what townships were burned over? Part lots ~~35~~ 34 Con. XIII Ballantyne, Part lots 1, 2, 3, 4, 5, 6, 7, 8, 9, 10, 11 12 Con. XIII Wilkes. Part lots 1, 2, 3, 4, 5, 6, 7, 8, 9, 10, 11, 12, 13 Con. XIV Wilkes Part lots 1, 2, 3, 4, 5, 6, 7, 8, 9, 10 Con. XII Wilkes
10. Number of acres of timbered land burned over 65 acres 1-C
    " " old cutting with some timber left 2535 2-C
    " " young growth (up to 6 inches average diameter) —
    " " barren or muskeg —
    " " grass or farm land —

Total number of acres burned over 2600

11. How much of this total area is (1) private land none (2) under license all (3) Crown land not under license none
12. State the commonest kind of tree on the area Hemlock
13. Number of acres of merchantable timber burned 65
14. What is your estimate of number of feet (or cords or ties, etc.) 150 Thousand feet Hemlock 13M feet Birch 15 Cords pulp
15. What private property was damaged? 1 sett logging camps $1000. 1000 Saw logs $2500 1 Railway Bridge $500 (NOTE.—Your statement of buildings, logs, pulpwood, ties, etc., with owners' names, must be based on personal inspection)
16. If unlicensed, is the area so located and of such a size that the timber can be put up for sale? —
17. Cost of this fire for outside help in fire fighting, $ Cost of fire paid by Fassett Lbr. Co.
18. Who authorized calling these men out? Fassett Lbr. Co.
19. Who was in charge of the fire-fighting crew? Ranger Tennant
20. Was this fire traceable to breaking any provision of the Fire Act? Yes
21. If so, what steps are being taken? Cost of fire paid by Fassett Lumber Co.

NOTICE: Map of burned-over area to be made out on back of this sheet marking lots, concessions and township name.

Ont. For. Br. Form 23—2000-5-23.

Pump #366 ran 35 hours at this fire

Copy to Dept. of Lands & Forests
Copy to J.R. Booth Ltd - Ottawa
Copy to Fassett Lumber Co. Fassett, Que.

Wm. H. Hawkins
Chief

W. J. Tennant
Ranger

promises to remedy matters, their operating permit was cancelled until some means of safer transportation could be evolved. This was effected by giving them a permit which stipulated that they haul only at night and that they must have a man located at their loader with a pump and hose. Subsequently the trouble from that quarter ceased and today the same company realizes that the forestry branch is a live organization with a function to perform and the power to fulfill it."

One of the new fire prevention measures implemented was a telephone line, strung through the trees from the Fassett **depot** on Tea Lake to the Lake Manitou Ranger Station. From the station it was connected to the main telephone system and Fossmill. This round about phone line allowed for business, as well as fire calls, to Fossmill. Of course, this did not prevent fires, but it shortened the response time to calls for help.

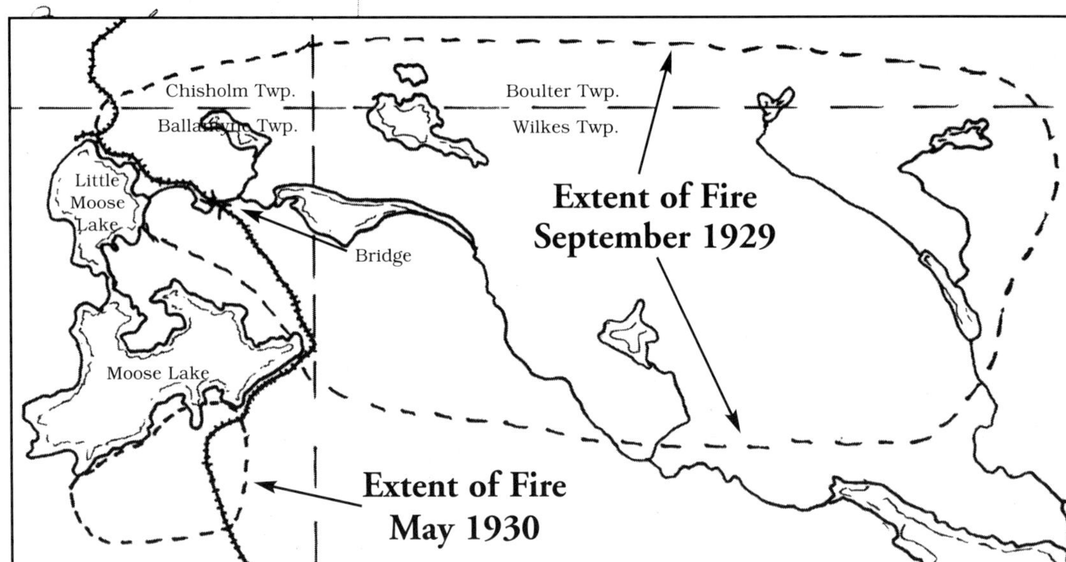

Continuing problems

"The railway frequently set little fires by ashes dumping out from the engine. We had to stop twice and put out a fire," said Park researcher Dr. Duncan MacLulich, who was taken into the park on the Fossmill railway. Once when he was coming out of the Park the lunch fire of the two company men who had come to meet him was out of control. "It was roaring right up to the tree tops," he said. As they fled the fire in their **gas car**, it collided with the speeding gas car of two rangers rushing to fight the fire. The fire rangers "didn't seem to get too much sympathy from the company men," Duncan said.

Earl Orr remembers a fire that came close to Fossmill. "I remember being in school. The fire was getting close because we could see the smoke and the flames out of the windows. The fences along the backs of the houses on the second row were burning. It was in the early spring because I remember Dad getting a team of horses and pulling the car off the blocks. We loaded the car and drove away from the fire. The wind shifted, and we were safe, but the hills were black. There were several forest fires around, but that was the one I really remember because it got so close."

The Great Depression

Along with the hardship caused by the forest fires, the residents of Fossmill also began to feel the effects of the Great Depression. Lumber sales had plummeted but, initially, the company continued to cut lumber as usual. Sales dropped to between 500,000 and 800,000 board feet per month, yet, until late

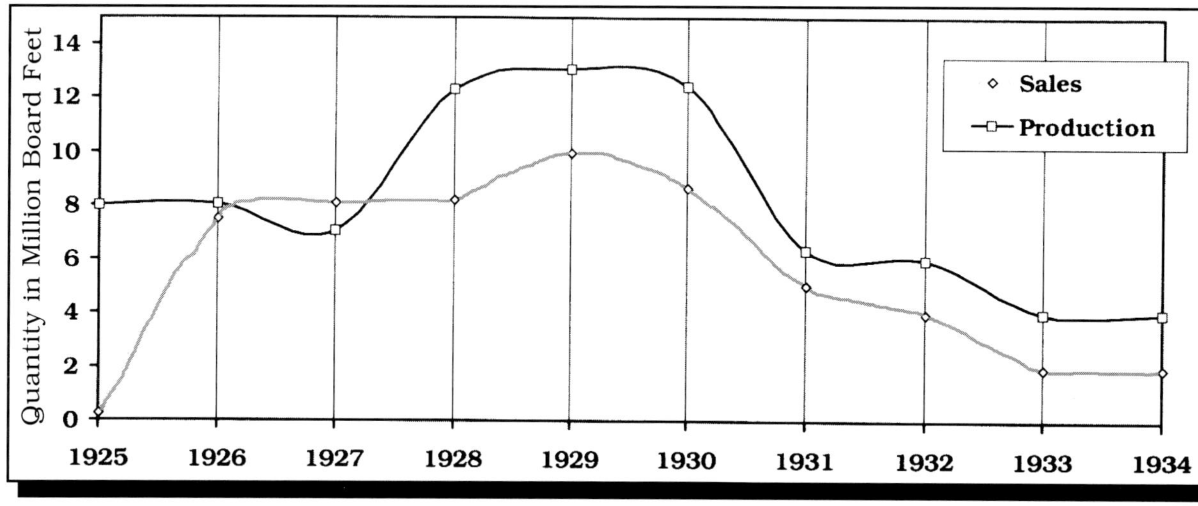

Over-production: *Statistics on sales and production during the Fossmill years show a policy of cutting much more than was sold. (figures from the last three years estimated).*

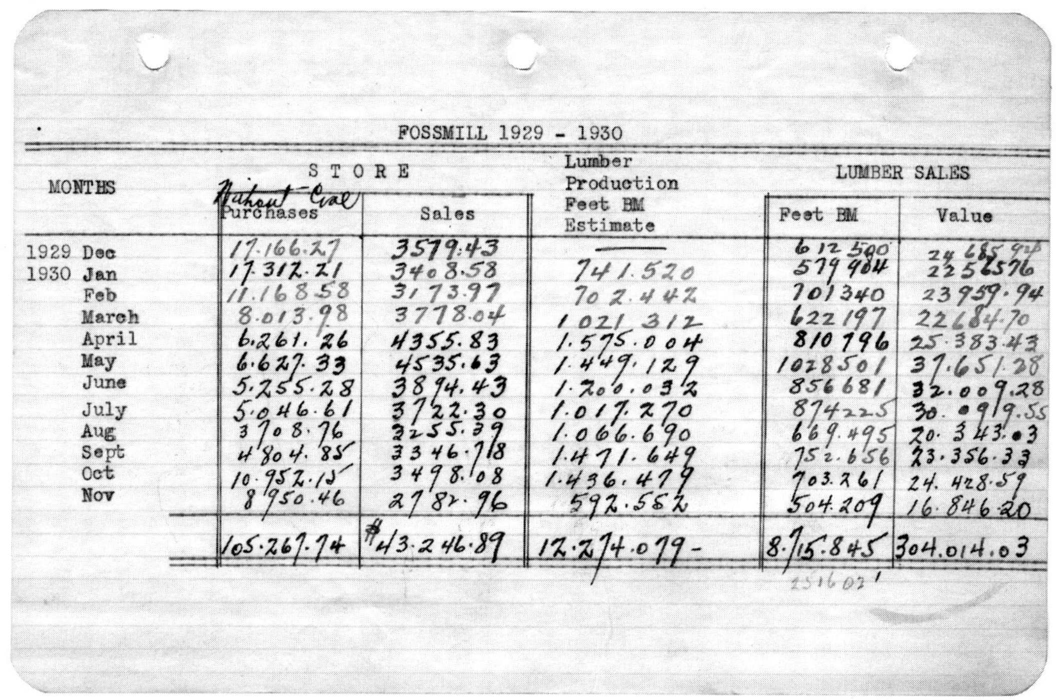

FOSSMILL 1929 - 1930

MONTHS	STORE Without Coal Purchases	STORE Sales	Lumber Production Feet BM Estimate	LUMBER SALES Feet BM	LUMBER SALES Value
1929 Dec	17.166.27	3579.43	—	612 500	24 685 92
1930 Jan	17.312.21	3408.58	741.520	579 984	22 565 76
Feb	11.168 58	3173.97	702.442	701340	23959.94
March	8.013.98	3778.04	1.021.312	622197	22684.70
April	6.261.26	4355.83	1.575.004	810796	25.383.43
May	6.627.33	4535.63	1.449.129	1028501	37.651.28
June	5.255.28	3894.43	1.700.032	856681	32.009.28
July	5.046.61	3722.30	1.017.270	874225	30.919.55
Aug	3708.76	2255.39	1.066.690	669.495	20.343.03
Sept	4804.85	3346.78	1.471.649	752.656	23.356.33
Oct	10.952.15	3498.08	1.436.479	703.261	24.428.59
Nov	8950.46	2782.96	592.552	504.209	16.846.20
	105.267.74	#43.246.89	12.274.079-	8.715.845	304.014.03

Records: *Sample page from Sydney Staniforth's personal record book.* [RS]

1930, 1 to 1.5 million board feet were cut per month. In November, 1930, production was reduced to 600,000 board feet per month. To raise cash and reduce costs the company sold at least one of its Cletrac tractors to Gillies Bros. Ltd. The mill was closed all of December 1930, and January and February of 1931. In February, Sydney Staniforth helped form the Hardwood Bureau, which began a campaign promoting Canadian hardwoods. Sales improved briefly in the spring of 1931 when the company sold just over one million board feet in May.

Berry picking

"A couple of years after the fire, the raspberry bushes just grew and grew. They were the most raspberries we had ever seen," says Margaret Gleason, remembering one positive outcome from the 1929/30 fires. As the Depression deepened, wages dropped and the mill closed temporarily. Wild berries were a low-cost source of good food in season and, when canned, for the long winter months. Margaret remembers that "The company said all those berries were up there going to waste. To help us out, they took anybody that wanted to go and pick raspberries in the caboose on the way to work. We were told to be down at the yard at half-past-five in the morning. We went to Little Moose most of the time. There was about 25 of us, including my brother and I. The dew was still on so we couldn't pick when we got there right away. We waited an hour or an hour and a half.

"We had our pails or baskets filled up before lunch—the berries were that good. I walked under the bushes and just picked, that's how high they were. In the afternoon we paddled around in the lake. They picked us up on their way home around four o'clock. We heard them coming so we had lots of time to get back to the railway. Mother would can the berries and make jam. We had good dug-out cellars. It was real cool down there so your fruit and stuff kept." Unfortunately it was on one those berry-picking expeditions that tragedy struck the community.

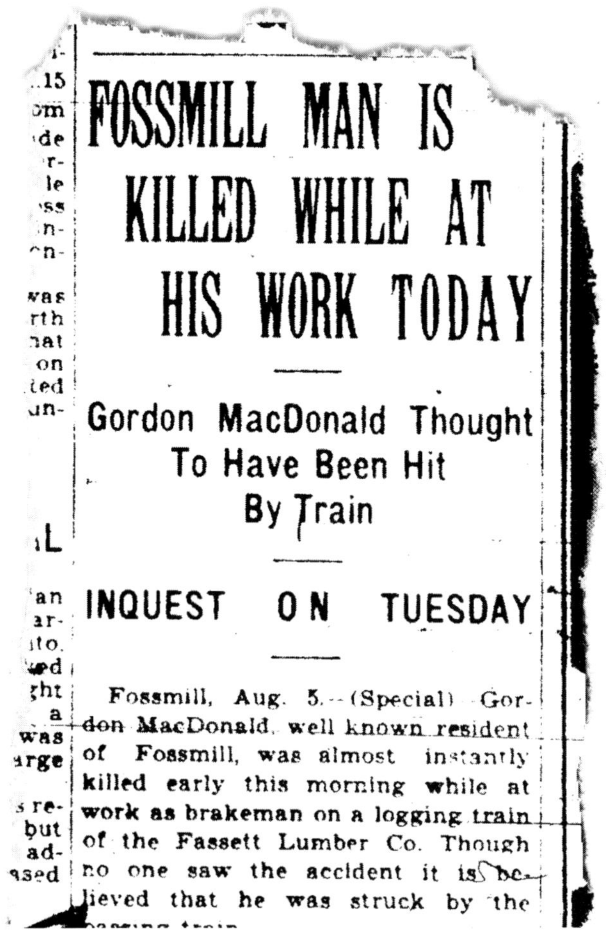

FOSSMILL MAN IS KILLED WHILE AT HIS WORK TODAY

Gordon MacDonald Thought To Have Been Hit By Train

INQUEST ON TUESDAY

Fossmill, Aug. 5.—(Special) Gordon MacDonald, well known resident of Fossmill, was almost instantly killed early this morning while at work as brakeman on a logging train of the Fassett Lumber Co. Though no one saw the accident it is believed that he was struck by the

Tragedy

Wednesday, August 5, 1931 started out as a pleasant day. Louis Demers was invited to join Gordon MacDonald's wife, Myrtle, who was pregnant with their 13th child, and their young son Freeman, on a berry picking excursion.

As usual, the locomotive went into the Park in reverse, pushing the log cars ahead. The caboose, containing Louis Demers and the MacDonalds, was coupled to the front of the locomotive and followed along behind. Myrtle pulled herself up into the **cupola** of the caboose and looked out the windows.

After winding their way into the Park they soon came to the old burn, and the fields of ripe berries. The train crew had decided to do a so-called **running shunt**, in which the caboose is uncoupled and transferred to the siding without stopping the train.(An illegal procedure on the main line railways.)

Why they decided to leave the caboose for the pickers that day is not known. Perhaps rain was threatening or maybe Gordon wanted to make it more comfortable for his wife. Germaine Demers recalls that "Myrtle didn't want to go that day, but he begged her to go."

As they approached the siding the conductor, Morris Donahue, stepped off the train and prepared to throw the switch to allow the caboose to go into the siding. Gordon swung up onto the front of the locomotive and pulled the pin holding the caboose. He then had to jump onto the

Clipping: *This article appeared in the* North Bay Nugget *on May 5, 1931.*

caboose to stop it when it was safely on the siding.

"As the caboose pulled away from the locomotive Gordon made the jump onto the caboose," remembers Louis Demers. "He fell backward between the brakes and the shaft for the wheel. We saw him go—he fell across the rails and was cut right in two. His wife was sitting up on top, she saw him look at her as he was going down."

Myrtle told the editors of *Our Chisholm Story* that Gordon's "foot became caught and apparently realizing nothing could save him, he gave Freeman [his son] a push which sent him back into the caboose, when he attempted to rescue him, as he was being drawn under the wheels."

Young Lionel Martin was in Fossmill that morning and remembers that the train returned to Fossmill "around 10 or 11 in the morning. They were parked down below the hill from the office. They stayed there for quite a while. Everybody was sitting on their porches looking in that direction. Then the men left and went back in the bush. They came back after dark with the day's quota of logs."

The *North Bay Nugget* reported that Gordon was "almost instantly killed" and that he had "suffered a broken neck, a fractured arm, and had bled to such an extent as would have caused death without other injuries." The same article noted that Gordon "was a highly respected and esteemed resident of the community. He is survived by his widow and 12 children besides his mother, six brothers and, one sister. He was a member of the United Church and of Chisholm Orange Lodge No. 1323"

Gordon's body was laid out in the upstairs bedroom of his house, and a wake was held to mourn his loss in the community. His daughter Myrtle, who was born shortly after his death, remembered stories about her father. "They said when there was any trouble at the mill they always called my father. When they had some other men bullying and picking on the younger men in the mill my father used to settle it. He called at all the square dances. He was crazy about children and children just loved him. He helped to build the church in Fossmill."

Even Father Gravelle attended Gordon's Protestant funeral. "He came to my father's funeral. The only thing was that when the Orangemen had their flags up, Father Gravelle walked around. The next Sunday he said Mass for my father," Myrtle said.

Although his death was an accident, the MacDonalds felt some bitterness. Myrtle recalled her mother's feelings. "My mom said my dad could never take a day off because they couldn't spare him, and after he was killed they had a man on the job right away."

Growing Depression

By the summer of 1931, the Depression was in full swing. Lumber sales dropped to a low of 250,000 board feet in August 1931. Even though the company had cut back production to only 750,000 board feet per month, lumber still piled up, and the yard was soon full to overflowing. Wages were reduced from a 1920s rate of around $3 a day, to only $2 a day. In the middle of August production was stopped and the workers were laid off.

"During the thirties—I remember them talking," says Donald Staniforth, who overheard his father's conversations with the other company owners. "There was no money. They couldn't sell the lumber, but they were running, and dad was giving the employees room and board, but couldn't pay them. It was just that tough. Nobody today could ever realize how tough it was—they couldn't sell the bloody lumber!" As the Depression deepened, the situation in the lumber industry was becoming worse. In September, the *Canada Lumberman* reported that "The hardwood situation...shows quite a decline in value and volume."

Without an income, some people took desperate measures to survive. One man remembers his father telling him how he got food for the family. "Dad used to go and play checkers at the store and he'd always ask the guy at the store for a couple of boxes. He'd say, 'There is some in the back,' and Dad would go in the back, grab a chunk of meat, throw it in one box and throw another box on top and walk out. He also used to play checkers by where they kept the eggs and he used to fill his pocket up with eggs."

Rachel and Amedee

Despite bad times, villagers tried to live their lives as best they could. In 1931, Rachel Denis was no longer required to take care of her family, and on September 1st she finally married her long time boyfriend Amedee Laferriere in the Fossmill Church. The joy of this occasion, however, was short lived.

Rachel Denis and Amedee Laferriere: *This photo is from around the time of their marriage in September 1931. [GL]*

September 30, 1931

On Wednesday September 30, 1931, workers were loading a boxcar with lumber, deep inside the Fossmill lumberyard when a fire was noticed in an adjacent lumber pile. The workers abandoned the boxcar, which was soon on fire, and sounded the alarm. The mill's steam whistle quickly brought the rest of the people in Fossmill and alerted the Fire Rangers.

"There was little fire protection. The company had a couple of barrels with water pails in them. There was no hydrant or anything," recalls Roger Bergeron. "As the fire spread into the yard, more and more wood piles burned. The heat and the flames were terrific."

The *North Bay Nugget* reported that the fire "was soon out of control and the efforts of the fire fighters were concentrated on the mill and the surrounding dwellings." Fire rangers and fire-fighting equipment were rushed in from Powassan and Brent, and a "man and pump" were sent from as far away as the Parry Sound District. Hoses were run to the Wasi creek and pumps were fired up.

The tramways from the mill to the lumber yard were cut to prevent the fire spreading to the mill. The *Nugget* reported that "By putting up a persistent battle, the men succeeded in keeping the flames in the nearby lumber piles from reaching the buildings, and for the remainder of the night several lines of hose were engaged in keeping the mill saturated with water."

The people who lived in houses close to the burning lumber piles were not so lucky. "There was a big coal storage shed right in front of our house," recalls Roger Bergeron. "They said, 'If we can keep the fire from getting to that coal pile we will probably be able to save your house.'"

His family had time to move their furniture well back from the blaze. They watched hoping for the best. Unfortunately, "the fire kept coming and it caught into the coal pile. The heat from it started the siding on the house, and boom—the house went up—burned right to the ground. My dad used to make dandelion and chokecherry wine. The wine was in gallon jugs up in the attic and nobody thought of it. After the fire got going you could hear them popping."

Engineer Dominic Boyer's house was just below the Bergeron house and it too met the same fate. Villagers feared for the whole town. Nora (Gleason) Clark, remembered seeing the fire from her bedroom window. "I was sick in bed. The CNR tracks passed by our front door. My father brought the little locomotive and a flat car up to take us out in case the whole place burned."

The *Nugget* reported, "When the fire was at its height, the reflection in the sky was visible for many miles around and could be seen plainly in North Bay, a distance of 40 miles away.…People from all over the surrounding district, including North Bay, were attracted by the reflection in the sky, and cars from many points lined the main street of the village."

"You couldn't get any closer than the railroad track because the heat was so bad it would burn the clothes right off of you," recalls Max Smith. "There were flames going 200 to 300 feet in the air. Everybody came. There were people from Powassan, North Bay and everywhere." Some people claimed that the fire could be seen 70 miles (113 km) away in Bracebridge.

It took 10 hours to get the fire under control, and it wasn't completely out for 48 hours. The *Canada Lumberman* reported that 6.5 million feet of high grade hardwood was destroyed, along with another 6 or 7 million feet of hemlock, spruce and white pine. Seven hundred cords of slabwood were also destroyed. It also reported that "All the trolley trucks and docks were consumed by the flames," and that "The steel rails were distorted into all kinds of shapes."

The Fire Report suggests that the fire may have been caused by a spark from the locomotive or a discarded cigarette, but the cause was reported as "unknown." There was little doubt in peoples' minds that it was an **incendiary fire**, set by someone to get the mill operating again. The company received $400,000 in insurance. The tramway and railway sidings were soon rebuilt, and the men were soon working again. The *Canada Lumberman* reported two months after the fire that "The mill was starting to cut 60,000 pieces and that Fassett was ready to put in several camps."

Homeless

Roger Bergeron's family was without a home after the fire, and were moved temporarily into the boarding house. "There was no house empty in Fossmill. They gave us rooms that were for the boarders," Roger says. "They were all individual rooms with bunk beds. We ate with the cook in the cookery after the men were finished eating. After a while a house became available in the village and we moved. I guess we stayed at the boarding house for the better part of a year."

Burning lumber: *Photograph of a fire in the Pembroke Lumber Co. yards in 1927, similar to the Fossmill fire.* [CL/RS]

Scarlet fever

Bad luck was coming to Fossmill from all directions. In November, the Desjardin, Denis, Trudell, and McGibbon families came down with scarlet fever and the Board of Health had to quarantine them.

"Half of one year my mother was sick," recalls Peggy McGibbon. "We ran through scarlet fever and pneumonia. Everyone was sick. We were quarantined for five of six weeks with scarlet fever. Nobody would come up there. My dad got a man named Mr. Cameron, who was a cook for the company and didn't mind coming. He did the cooking and cleaning and looked after us." As a result of the scarlet fever Stella got an ear infection that affected her for the rest of her life.

The pressure of the Depression on manager Tom Howard, who was trying to handle the situation, and the pressure on the community as a result of death, fire, unemployment and sickness, made Christmas 1931 a trying time for the community.

As well, Boyd MacDonald was back in Fossmill. His father was dead and his mother was having a difficult pregnancy. Doctors thought she had a tumour growing alongside Myrtle, her 13th child.

Midnight Mass

On Christmas Eve some of the Catholics in Fossmill wanted to attend the Midnight Mass that Father Gravelle was holding at the Saint Louis de France church in Chiswick. Someone suggested that they borrow a team of horses and a sleigh from the company stables. Boyd MacDonald, always ready to help out, went to the stable boss and asked to borrow a team. He was refused. The *North Bay Nugget*

Rebuilding: *In 1931, Tramways were quickly rebuilt over the ashes and twisted rails of the old tramways and spur lines.* [DL]

In the news again: *Article from the* North Bay Nugget, *January, 1932.*

...Public Works Com-
...tee to right such wrongs. "Well, I certainly will do it," he said.

YOUTH IS SENT TO JAIL FOR TAKING HORSES

Foss Mill Lad Used Horses Without Permission to Drive to Midnight Mass

Boyd McDonald, Fossmill, was sentenced to one month in the district jail by Magistrate McGaughey on Tuesday afternoon when he pleaded guilty to having broken and entered the stables of the Fassett Lumber Company on Christmas eve, and taken out a team of horses which he drove to midnight mass at Chiswick. Ambrose Dwyer also pleaded guilty to the same charge and was allowed to go on suspended sentence of one year. Two others charged with the same offence pleaded not guilty and the charges against them were dismissed.

McDonald, during the evening of Dec. 24, had asked the stable boss for the loan of the team and had been refused. He later, in company with other young ...

reported that later that evening Boyd, "in the company with other young men, went to the stable. MacDonald entered the building through a window and then admitted the others through a door which he opened." The crowd piled on and Boyd took them to Mass.

"At 5:30 on Christmas morning the stable boss found the team in their stalls but showing signs of having been driven," reported the *Nugget.* Tom Howard called the police, and Boyd and three other boys were charged with stealing the horses. A comment by one of Boyd's group provides a hint at a reason for such drastic measures. He said, "We were a bad bunch of young fellas. He [Tom Howard] was pretty bad at times—he wasn't too friendly—he wanted us out of there. We had nothing to do, no work or anything."

At the trial Boyd, and Ambrose "Mike" Dwyer, son of the sawfiler, pled guilty to breaking and entering. Boyd received one month in the district jail and Ambrose received a one-year suspended sentence. The other two boys pled not guilty and the charges against them were dismissed. The *Nugget* reported that Boyd "is said to have told them that he had permission to take the horses." Some people say that Boyd was a "fall guy" and that Tom Howard's action was a "bad mistake."

The outcome of the problems with Boyd marked the MacDonald family for years. Boyd's sister, Myrtle, said later that "Tom Howard didn't like Boyd. My mom said that he came and told her it's either you move out of Fossmill with all your family or send him away—well, how could she move out of Fossmill with the children she had?" Around this time, Boyd's sister, Elva, married Lawrence Huneault. "Lawrence didn't want to stay in town because he didn't like old Tom Howard," recalls Lawrence's brother Donat. Lawrence and Elva built a log house east of Fossmill, just outside company property.

In contrast, the community pulled together that New Year when Mrs. MacDonald had the difficult birth of her namesake, Myrtle. It was New Year's Eve. Doctor Dillane from Powassan was away, so his replacement Dr. White was called. Apparently Dr. White was going to a party with his girlfriend and arrived with her in tow. Myrtle told the story of her birth. "When I was born my mother had a terrible time because of the tumour she had. The tumour came out with me so she was very, very sick. Father Gravelle used to come and sit with her. Lawrence's mother, Mémère Huneault, and Aunt Maggie Donahue helped. They all came and looked after my mom."

Mrs. MacDonald eventually did get some Workers' Compensation benefits from the government for the death of her husband. She continued to live in Fossmill with her children for about two years and eventually moved to a house in East Ferris. Myrtle recalled that her

MacDonald clan: *Mrs. Myrtle MacDonald (centre) held the family together after the death of her husband.* [MC]

"mom had to move out of Fossmill. They told her that she had to go, there was nobody from her family working there, and she was taking up a company house. I think we moved out of there when I was two— they wanted the house for somebody else."

Military intelligence ?

In 1941, Boyd MacDonald, along with four of his brothers and a brother-in-law, signed up to fight in the Second World War.

Private: *Boyd Macdonald in 1940 after joining the army. He was member of the Canadian Provost Corps. [NBN]*

His mother received a "mother's medal" and was the subject of an article in the *North Bay Nugget.*

Boyd was, as we have seen, a defender of the underdog, but, in the wisdom of the military, he was assigned to the military police and sent to pick up draft-dodgers in Quebec! "They did that because he was a scrapper and he used to box," sister Myrtle recalled.

It is no wonder that Boyd went AWOL and returned to the family home in North Bay. Myrtle remembers that "He'd hide in the trees outside the house. They were looking for him. We'd see the MPs' silver buttons shining when we went to the outhouse at night. All these MPs were out there waiting to catch him, and he'd be up in a tree or he'd be on the fire escape at the school just down the road."

Boyd was eventually discharged from the army, married and had three children. He died in 1968 in North Bay.

Back in Fossmill

In the winter of 1931-32 the lumberyard was empty, so, to fill it Fassett put in several camps with the insurance money from the fire, and hired some of the Fossmill workers. Rachel recalled that her new husband, Amedee Laferriere, worked as a cookee in one of the camps. At home in Fossmill he was the barber, cutting the hair of workers from the mill and the logging camps. "He got 50¢ from the guys from the bush. Sometimes they were six months without a haircut, and the smell of it, I could smell it in our spare room after they left," Rachel said.

That winter, other lumber companies in Algonquin Park were struggling. The Gillies Brothers did not operate at all. The McRae, Booth, and Armstrong companies operated "on a more or less reduced scale," reported the Park superintendent. In the winter of 1932-33 the Fassett Lumber company was the only company logging inside of the Park.

Inside the Park

Unlike Fassett who only logged in the Park, some of the other lumber companies had built their mills inside the Park itself. The Park superintendent was becoming concerned that existence of these mills brought "into being a small village and when the mill fails, these people have no means of moving out, and become another problem for the Park administration." In his 1932 report he wrote that "the chaos resulting from bankrupt logging operations in the Park has stranded several dozen families….As soon as the depression lifts, every effort should be made to induce these people to leave the Park. They are a nuisance and each community needs watching by rangers."

Hard times

The situation was critical for communities outside the Park as well. Each town or township was responsible for raising relief payments for the people in their jurisdiction, but unemployment was so high that the local governments could not pay. By 1933 the Federal and Provincial governments had assumed two thirds of the cost, but the one third still remaining was still a burden for the local government. That year Chisholm

Township reported 357 people on relief and notified the government that they could "no longer furnish relief, as the treasury is bare."

The Fassett Lumber Corporation struggled to survive. The Depression reduced sales to Fassett's traditional American buyers, so Sydney Staniforth actively went after new overseas markets. Articles in the *Canada Lumberman* in the early 1930s reported on the increased efforts by the Canadian Hardwood Bureau (Sydney Staniforth was the President) to promote lumber in Britain, and contained photos of displays at trade shows featuring automobile frames and other products made from Canadian hardwood.

Considering the fate of many other communities, Fossmill was weathering the Depression reasonably well. In the fall of 1933 Fassett put in two camps, employing 145 people—a far cry from the 400 it had employed in the 1920s, but at least they were hiring locally rather than bringing in outside jobbers. And 1934 began on an optimistic note—the Fossmill workers became movie stars.

Promoting hardwood: *A display for the Canadian Hardwood Bureau emphasizing the use of hardwood for various products including the automobile frame (centre).* [CL/UTL]

Hard times: *The Fassett Lumber Corporation was the only lumber company cutting in Algonquin Park in the winter of 1932-33.* [DS]

Earl & Ruth Orr: *With aunt Areta (Olimer) and Tom Cossar in front of the Orr house in Fossmill village. [JD]*

9

1934

"There was nothing anybody could do—we just stood there and watched."

In January 1934, a local newspaper announced that "An English moving picture corporation is taking pictures of lumbering operations of the Fassett Lumber Corporation at Fossmill." Sydney Staniforth had commissioned the movie as an advertisement for the company. Donat Huneault, a mill worker at the time, remembers the excitement this created.

Two filmmakers came to Fossmill. Donat remembers that the cameraman had a strong English accent. The other man was the producer/director. He picked the best quality lumber and largest logs for the various scenes in the film. Donat recalls that the film was made for the British lumber buyers so that they "could see that Canadian lumbering was something special" and how "tough the guys were." He recalls that the workers hammed it up for the camera and made the work look harder than it was. After the excitement of movie making, things returned to normal in Fossmill, for a while.

Strike in South River

In April, news arrived of labour trouble in South River, where Tom Howard's brother managed the Standard Chemical Company. The *North Bay Nugget* reported on April 23rd that "Finnish wood cutters are meeting here this afternoon to organize a branch of the Woods Workers Union. Delegates are here from Sprucedale and surrounding district. It is the intention to demand a rate of \$2.25 per cord for cutting cordwood instead of the current rate of \$1.25. All Finnish workers employed by the Standard Chemical Company quit work this morning, refusing to continue cutting at the present rate of pay. These have been joined by

"Trespasser": *An unemployed man climbs aboard a CNR freight train.* *[GA#NC--6-12955-(b)]*

others employed by jobbers."

The strike promised to become violent when new workers were hired by the jobbers. The *Nugget* reported that there were "some disturbances in the early stages of the strike when workers brought in by contractors were threatened." In August, the strike proved successful when workers received workers' compensation coverage for injuries and a rate of $1.75 per cord for cutting cordwood.

The "Quints"

On May 28th, 1934, the people of Fossmill heard that one of their neighbours, Mrs. Elzire Dionne in nearby Corbeil, had given birth to five identical daughters. The birth and survival of the babies would make headlines around the world and capture the imagination of millions. In the depths of the Depression this story would help take peoples' minds off the unemployment and hardship they faced. The residents of Fossmill still struggled to get by on the little work the company provided. In the rest of the country thousands of families were on relief and thousands of young men continued to "ride the rods" looking for work.

June 2, 1934

One of these "riders" was Frank Karlsson. No one knows whether he was attempting to return home to Vancouver when he climbed aboard a CNR freight train that was heading west through Algonquin Park that Saturday morning, June 2, 1934. He avoided the boxcar marked "explosives", and, because it was a clear day, joined the other dozen or so "hobos" or "trespassers" who were "taking free rides" on the open coal cars that followed behind the boxcars.

Many villagers remember when the train rounded the bend into Fossmill later that afternoon. Margaret Gleason recalls, "My girlfriend and I were going down to the store and the train was coming. We stopped to count the cars. You know how teenagers are. There was nothing much to do in a little village like that." Roger Bergeron was walking to the store with his sister. Earl Orr was playing in the school yard and Lionel Martin remembers that he was watching the train go by. Marie Danis' house was right beside the railway embankment. As she worked away in her kitchen, she needed something from her root cellar and went down the stairs into the room under her house.

Station agent Bill Casselman was standing at the window of the station as the train approached. "He thought he saw some strange movement of a car so he looked farther back," recalls Peggy McGibbon. "Another car was doing the same thing. He took a third look and realized a wheel was off the track. He jumped out of the back window of the station, twelve feet to the ground." Margaret and her friend watched in shock and surprise. She remembers that "All at once this boxcar went up into the air and next thing there was a black cloud. When that cleared away there was all this pileup." Roger remembers the "big cloud of smoke and dust—you couldn't see anything! After it cleared the cars were tumbled all over the place."

"The first car left the track at a switch about 50 feet west of the Fossmill station. The switch was demolished," reported The *North Bay Nugget.* A boxcar standing on the company siding was driven from the tracks, and derailed coal cars slid down the embankment toward the houses beside the railway. In total, 12 cars left the track and the rails were churned up and bent. The *Nugget* reported that the "residents of Fossmill said the noise was terrific as the crash occurred. The cars were badly wrecked as they piled up, wheels being torn off." Earl Orr, who was in the school yard, remembers the experience differently. He says, "It was funny. There was hardly any sound, but you could feel the concussion." Marie Danis emerged from her root cellar after hearing all the commotion and wondering what was happening. Her grandson, Raymond, recalls that a rail had pierced her house, coming in one corner and out the other. "She could have been dead," he says.

Many of the men riding the cars "hurried for the bush" and "there were stories of others digging themselves out of the coal, black with coal dust, but lucky to be alive," said the *Nugget*.

People rushing to the scene soon realized

Tumbled: *Workers examine derailed boxcars in Fossmill train wreck.* [CWI]

Wreckage: *Frank Karlsson's body was found under the devastation.* [CWI]

Marie Danis' House: *Wrecked cars came close to the houses beside the tracks.* *[LG]*

Main line: *Overturned cars were quickly pushed off the tracks to get the trains running again.* *[CWI]*

that a more serious threat lay in the wreckage. The *Nugget* reported that "One of the wrecked cars was loaded with…highly explosive dynamite," and that "Broken dynamite was scattered for 50 yards along the track." As people gathered around the scene, fire broke out under the dynamite car, and everyone was told to get as far away as possible from the wreck. "Quick work on the part of one of the company employees extinguished the blaze" and averted what might have been a devastating explosion.

Frank Karlsson was not so lucky. The *Nugget* reported that Karlsson "was pinned beneath one of the wrecked cars, and badly mangled." They speculated that there was still some coal to be removed from the side track, "in which bodies may be buried," but no new bodies were found.

The CNR rushed in breakdown trains, with big cranes, from Capreol and Ottawa and quickly cleared the track. On June 4th, the *Nugget* reported that "Frank Karlsson will be buried today, the funeral service taking place from Martyn's parlors."

The train wreck at Fossmill is legendary in Chisholm. Many believe that several hobos were killed. Children are still frightened by tales of the ghost of Fossmill, a hobo buried without his hand, who returns to look for it. Fran Gleason writes, "For a long time after the wreck, I had nightmares about these men coming up out of the coal, covered in black dust, knocking on our door wanting something to eat." Fossmill was lucky that day. The town was spared and none of their own was killed.

The summer of 1934 proved to be as eventful as the early part of the year when a hot summer was interrupted by snow. Doreen Smith reported in *Our Chisholm Story* that "A freak snow storm in July, 1934, enabled the men to bring home a sleigh, on the snow." The next day the heat returned and August turned out to be a sweltering month.

The good news was that there were some signs that the lumber industry was improving. Lumber companies in the North Bay area were planning on putting in 11 camps and hiring 1,395 men for the next season. The bad news was that the limits south of Tea Lake in Wilkes Township that Fassett had coveted since 1924, were auctioned off by the government and had gone to the Gillies Bros. Ltd. lumber company.

Day of rest

Some people remember stories of a "strange guy" who started working in the mill that August. Another remembers his grandfather saying that he was directed to do some unusual maintenance work that weekend. Everyone remembers where they were that Sunday August 26, 1934. "I was in the store with my mother," remembers Jerry Donahue. "We were coming back from Wasi Lake," remembers Roger Bergeron, who was returning to Fossmill in his father's car with his father and uncle Albert after a day of pickerel fishing. Margaret Gleason remembers the baseball game at the diamond just outside of town.

Breakdown trains: *A coal car is turned over. The CNR brought in two mobile steam cranes to clear the wreckage.*
[RO&CWI]

At 5:15 that afternoon, seven-year-old Lionel Martin was playing outside his house while his father did some chores inside. He was looking across the field toward the hot pond and mill when he saw flames "breaking through" the window beside the log deck and carriage on the second floor of the mill. "I told my dad. He looked out and started running towards the mill. I went over to Levis' and I yelled to them that the mill was on fire. Mr. Levis' son Roland came to the door and said, 'Huh? The mill on fire? What are you talking about?' Then Mr. Levis came bursting out and everybody came bursting out." Jerry Donahue recalls that the store had "a rear door they used to back trucks up to and I remember my dad running out that door towards the mill."

Hilaire Bedard, the lumberyard foreman, and his son also noticed the fire. They ran into the mill office and pulled the cord that set off the mill's steam whistle. As they ran from the office Hilaire decided to return and retrieve the time books. Smoke surrounded him and he was almost overcome. His son dragged him to safety. George Logan, the assistant stationary engineer, is credited with marshalling the men and starting to fight the fire. Nevertheless, within 15 minutes, the fire was out of control.

Roger Bergeron's day of fishing was quickly forgotten. He says, "On the way back we could see smoke. My dad said, 'Don't tell me that the yard is on fire again.'" Perhaps they slowed down to pass the baseball players running toward Fossmill and then sped off toward the mill. "When we arrived my dad said, 'all my tools are in there.' There were

The fire: *The mill was levelled in 45 minutes and spread to the engine house burning the loader and locomotive No 51.(left)* [RD/L/LD/CWI]

three great big boxes of every kind of tool imaginable, everything went and not one cent of insurance on it. There was nothing anybody could do—we just stood there and watched." The *North Bay Nugget* reported that the "flames leaped to terrific heights, well over the top of the mill's 100-foot smoke stack…a strong wind fanned the blaze."

Within 45 minutes the mill was leveled and the fire was spreading. They realized that the engine house, lumberyard, boxcars, outbuildings, and the town itself were in danger. "It was only the Trojan-like efforts of several hundred men and women that saved the lumberyard," reported the *Nugget.* Boxcars were moved out of the way. Three houses close to the mill caught fire but "prompt work saved the structures." Fire rangers came in from Fossmill, Powassan, and Brent. The CNR sent in a special fire fighting unit.

The engine house and blacksmith shop could not be saved. Locomotive No. 51, one of the log loaders, and some log cars were destroyed in the inferno. The *Journal* reported that "A light rain about 9 o'clock helped the firefighters, but men worked in darkness, pierced only by the flames, as the town's power plant was destroyed in the fire." The fire ranger's report stated that the fire was under control by 1:00 a.m. on August 27th and that fire fighting ended at noon. Early estimates of damage ranged from $170,000 to $250,000, not including the valuable tools of Alderic Bergeron, John Orr, Ludger Levis, Percy Dwyer, Joseph Sliwka and others. The

Loader: *The melted and tangled remains of the loader are on a flatcar. The remains of Locomotive No. 51 are behind the caboose. [DS]*

Nugget reported that "Insurance is carried on the mill, but about 100 men will be thrown out of work. The mill was the sustenance of the town….What the company's future plans are is unknown."

The fire ranger's report listed the cause of the fire as "unknown." The *Nugget* reported, "The fire is believed to have started in the lath mill, and within a few minutes spread through the entire building." Some workers remembered the "smell of kerosene," and that the stranger who had recently started work "disappeared."

Twisted: *Only the iron and steel were left after the fire. The village, uncertain of its future, is in the background. [DS]*

The cause of the fire was never determined. "The rumours at that time were that the fires were set," recalls Roger Bergeron. "There is no proof, but the rumours were that somebody had been paid to do it—then again maybe it's not true. The timbers in the mill where the machines were running were all soaked with oil and 'boom'—just like that—and no protection. What have you got for fire protection? Nothing whatsoever."

Sad sight: *Charred remains of the "little engine" No. 51. [DS]*

Rebuild the mill?

The lumberyard was saved. It had 8 or 9 million feet of lumber in it, and 24,000 softwood logs were piled along the railway. Would the company rebuild the mill? Many thought that it would not, because the limits in the Park were believed to have been close to exhaustion.

One villager said, "As far as I can remember, talking to my Dad, it was sort of a 'fait accompli' when it burned. They weren't going to rebuild it because they were sort of running out of areas in the Park where they could get permits to cut." Another said, "They were through. There were no more logs up there; they were ready to move on. This is what they do with their mills, isn't it?"

Peggy McGibbon disagrees and feels that the fires were a result of "the Depression. The timber was still there. My father's original estimate was right." In 1928, The forestry students estimated that there were 96 million board feet of timber on the limit. Company production figures show that by 1934 they had cut around 79 million feet leaving 17 million feet in the bush. But was it profitable for the company to rebuild the mill? They had invested heavily in the railway, town, and in building a skilled workforce. The future of

Clean up: *The remaining loader and the "big engine", Locomotive No. 54. tidy up the area in front of the boarding house and company store. [DS]*

Fossmill rested on the business decision of the owners.

One day, shortly after the fire, Sydney Staniforth and Charles Dale, after looking over the destruction, held a meeting in the company office above the store. Anxious villagers gathered to hear the verdict.

"There was a meeting, up in the office, between Dale and Staniforth," Lionel Martin recalls. "The windows were open and they were talking very loud. Staniforth said we will rebuild the mill and Dale said no we are not rebuilding. All the workers were listening outside." Many workers "still believed, for years after, that it was Dale's fault they didn't rebuild in Fossmill." Sydney Staniforth's son remembers that "After the mill burned they decided to part company....I recall that there was a fight, a misunderstanding, with Dale."

As Sydney left that meeting it must have been hard for him to look at the people outside. Sydney was offered a management position in the United States and began commuting from Montreal to North Stratford, New Hampshire. But he was not about to let his knowledge, connections, and the experienced workers at Fossmill go to waste. He had his eye on a spot 14 miles (23 km) east of Fossmill situated beside the CNR, on the shores of Lake Kioshkokowi inside Algonquin Park.

The future of the community in Fossmill was uncertain. Fran Gleason wrote that "The lone smokestack stood where the mill had been. I remember the men cutting the guy wires that held it up. One was on the far side of Wasi Creek and another on the opposite hillside. When that stack fell it seemed to mark the end of Fossmill and life as we had known it."

Firewood: *Orien Demers and Lucian Danis cut up old timbers for firewood.* [RD]

10

ON RELIEF

"I remember my mom crying lots of times."

"I remember the noise and smell of the little mill—I love the wood smell," says Barbara Scott (Hendrickson). "I remember going into the mill one time taking a telegram at lunch, and the food, the food they were eating, you know, pies and cakes and everything."

Jack Scott, Barbara's father, became the new station agent at Fossmill soon after the mill fire. One of his jobs was to receive telegrams at the station office. Barbara remembers, "It had a window in front and that was where the telegraph key was. At one time I learned the Morse Code."

The Fassett Lumber Corporation, although in disarray, continued to clean up what remained after the fire, while the villagers considered the limited opportunities available to them in Fossmill. Jack McGibbon's family moved to Powassan, leaving their house available for the new station agent. John Orr, the stationary

Bergerons: *Albert and Aurea Bergeron with children Robert and Lucille. The Martin's house (formerly the MacDonald's) is on the right in the background.* [LG]

engineer, moved to Callander to other work. But there were still millions of feet of lumber in the yard to sell, and 24,000 logs waited at the dumps by the railway. Tom Howard salvaged the mill's boiler and built a makeshift mill in the ashes of the old one. It was to this mill and to Tom Howard that Barbara took the telegrams.

The men eating the pies and cakes were the few men who were able to secure jobs in the little mill. Allan Gleason, for one, became the engineer in the mill. Tom Howard was in charge of this reduced operation. He commuted to Fossmill from North Bay during

Temporary mill: *Framing for the new temporary mill goes up in the background. The steam power plant is rebuilt in the foreground.* [PL]

the week. "I got the idea that he came and went a lot. There was some kind of apartment or rooms that he stayed in above the store," Barbara recalls.

The boarding house had burned down in another fire in July 1935, but Tom Howard continued to operate the company store. Although most people did not work for the company anymore, and some people were moving in from outside, the company still owned all the houses in the village. It was Tom's job to collect the rents from the villagers. Roger Bergeron remembers that the company "ran for a couple of years. There were all the houses there. Nobody owned the houses. They all paid their rent to Mr. Howard."

Relief

Aside from men working in the mill the rest of the residents of Fossmill were left to fend for themselves. It was the middle of the Depression and Canadian unemployment was around 14%. Without employment, most families had to go on Relief. They could not move out of Chisholm Township to look for a job, because most jurisdictions required recipients of relief payments to have lived locally for a number of years.

Relief payments were around $5 to $20 per month for a family. Recipients were required to turn in car licenses, prove they had no food in the house, and work on various public works projects. "We had to work for those payments. We used to walk as far as Alderdale to cut brush and widen the roads for the council," remembers Donat Huneault. Roger

Relief roadwork: *Men on Relief work on road in Haliburton County. [AO#AU-2513-S101]*

Bergeron says, "My dad went out to work after the mill burned down in '34. He used to work on the stretch of road around Wasing."

Raymond Danis remembers his father cutting wood to be sent to the old North Bay jail where it was distributed to people on Relief in North Bay. Peggy McGibbon recalls that during the Depression "there was no place to go. They couldn't get any work. The young men from Fossmill worked on constructing Highway 17 from North Bay to Mattawa for $5 a month."

The small relief payments were never received in cash. "We never saw the colour of a dollar bill," says Roger Bergeron. "The township would issue you a voucher. The voucher was only good for food and clothing. We used to get all our food from Butler's store. You would go in with a voucher. If you had a $30 voucher and if you took $20 worth of food he gave you back a $10 voucher. You never saw cash."

The Fossmill store also took vouchers. An invoice from the Fossmill store to the relief officer for Chisholm township in early 1936 lists 21 families on Relief in Fossmill.(see Appendix B page 209)

A few people found work. "I was only 15 when I had to leave Fossmill to go to work, because there was no money—it was bad," says Germaine Demers, who worked as a maid. "I thought I better leave and go to work. Dad still had two youngsters at home."

A job: *Temporary firefighters take a break while fighting a fire near Parry Sound.*
[AO#C7-13509-(A-421)]

Fire fighting

Forest fires provided work as fire fighters for many men. Many fires were deliberately set to make work and were clearly reported as "incendiary." The report on a fire in Wilkes Township stated that there was "no proof of identity of guilty parties, though satisfied fires are incendiary." The report on another fire stated, "Fire fighters passed close to the point where fire started twice on August 31 and September 1, and there was no fire. On going to work September 2 this fire was burning, leading to supposition that fire was incendiary." Other fire-reports had similar comments.

When John Gleason was 14 years old he worked for the local fire rangers driving them around in the family car. "One summer I used to drive for them. I drove everywhere. We used to go to Mattawa, up as far as North Bay and down to South River. They had that rectangle. I got paid mileage for my dad's car and 20¢ an hour." John also remembers that during that summer, "times were tough—no doubt about that. They would hire everybody they could. The workers also got 20¢ an hour. There were fires everywhere north of Fossmill, small poplar and stuff like that. I drew maps for them. The ranger sat in his car all day. I walked for miles and made maps for his reports. You would see these guys ahead starting fires and the guys behind putting them out, to make work. There wasn't any work anywhere else. I would go to the store in Alderdale and load up with cans of beans and bologna for the men. They worked all night sometimes."

Self-employment

Other people tried various projects to survive. Morris Donahue's son, Raymond, tried to start his own business. "Everything went wrong for Raymond. He started a chicken farm in Fossmill and one night one of the heaters caught fire and burned the whole thing down—all the little chicks died," recalls his nephew Jerry Donahue.

Young John Rouselle also tried his hand at self-employment. "I would go and catch rabbits with snares and stuff like that. You wouldn't believe it. I would catch, sometimes, 20 a day. I used to skin them and when the little train came in I would sell them for 25¢ a rabbit."

"The Mixed"

During this period most of the cars in town were either sold or could not be used, so the local train became the link with the outside world. "We'd go down and meet every train, that was a big occasion," remembers Barbara Scott. "They didn't have cars, I don't remember cars in town at all."

The train was referred to as "The Mixed" because it carried both passengers and local freight. John Rouselle recalls that it was "just a short train, two or three boxcars, a coach, caboose, and an engine. It came from Capreol. It picked up stuff like apples, meat and this and that; it picked you up too. The conductor charged you 50¢. They came down the main line and stopped at Alderdale, Fossmill, Kilrush, Kiosk, Brent, all the way down. It turned around the next morning and came back. Some people hopped on and went to North Bay to shop."

Shopping

The luxury of shopping in North Bay was reserved for better times, since most did not have the money for train fare. The company store provided some goods, as did the general store in Alderdale. Travelling salesmen also came by. Barbara recalls their visits to Fossmill. "A man came with meat products. I think he had horses because the only cars I remember were the Rawleigh and the Fuller Brush people. I remember buying stuff at the door because you couldn't get out to shop anywhere. Mother was quite a cook, so she used to buy spices from the Rawleigh people."

Without cars or train fare, walking was often the only option. Ila Huneault, the daughter of Lawrence and Elva Huneault, was born in a hospital in North Bay in 1935. "When I was in the hospital my dad walked Alderdale Road to North Bay to see me and back again to Fossmill." Her homecoming was to the log house Lawrence and Elva built outside Fossmill. "When we got off the train in Fossmill, my mom and dad brought me home by dog sled."

Hobos

The trains also brought many of the men who were crossing the country trying to find work. Some found a temporary welcome in Fossmill. Barbara Scott remembers that "There were lots of hobos and my parents never turned them away." Her father let them sleep in his station waiting room and her mother gave them some work. "Mother had a garden and we always had huge stews. At least two or three days a week we would have somebody to dinner. Mother always made them do something. They'd chop wood or maybe dig the garden—for pride's sake—I guess. It always seemed so sad to me. One man in particular, I remember, kept talking about his children that he hadn't seen for a long time. To me it was a very, sort of, sad time, between the village not having enough and the hobos."

Some of the hobos would try to take up residence in town and regrettably had to be moved on. "I think some of them tried to move into empty sheds and things and stay and they kept pushing them out. I remember hearing people say they were down in some shed or other," Barbara says.

A woman from Eau Claire told the story of a visit she made to Fossmill and her experience with hobos when she was a child during this era. One evening, when the train was in at the station, there were a couple of hobos on board. She took her friend's dog for a walk along the track. As the train pulled out one of the men jumped off, scooped up the dog and jumped back onto the train. The girl stood there in shock as the train pulled away taking the hobo and the dog with it.

Waiting for a ride: *Hobos and unemployed men looking for work and hand outs were common in Fossmill during the Depression.* *[GA#NC-6-12955-(i)]*

School

After the mill fire in 1934 the Fossmill school continued as an English public school although most of the students spoke French. The new teacher was Gordon Gourlay. He tried to satisfy both the English-speaking and the French-speaking communities; however, members of both were becoming dissatisfied.

"I went to the public school all along," Roger Bergeron recalls. "Then Father Gravelle in the last two years said to me, 'You are going to come to school in Chiswick. You are getting no French at all.' I had a CCM bicycle and I used to bicycle from Fossmill to the school in Chiswick right beside the church. There were a couple of nuns teaching there."

"I only went to the Fossmill school for a few months," recalls Barbara Scott. "The kids were all French. There were hardly any English-speaking kids. I was just kind of left on the side, so my parents decided that I should go to the Wasing school. My father had been a public school teacher before he joined the railway, and he decided I wasn't getting a good enough education."

Church

The Catholic Church in Fossmill became a centre of spiritual support in those hard times. The spiritual nourishment was led by "Aunt Maggie" Donahue, sister of Morris Donahue, the logging railway superintendent. "Aunt Maggie was always the one," recalls one villager. "We used to go to the church for prayers. She always led the prayers. She was so precise and slow and she would do it like you are supposed to do it. And then us kids would jump in and ring them off."

Many non-Catholic children also attended Catholic services. "My mother was down shopping one day," recalls Barbara Scott. "A lady said to my mother 'it's wonderful that Barbara is going to be one of the saved,' and mother sort of looked at her. What had happened you see, we went to the Wasing Church, but church was never till the afternoon, so I went to the Catholic Church every morning with all my friends—still in the back of my head I can say some of the catechisms—we were drilled every day and, of course, I wanted to be with my friends. There were 12 to 15 kids—everyone my age went. I can remember mother laughing about that—I went on my own. They didn't know where I was. I was supposed to be playing with my girlfriends." She did say her parents didn't object. "We moved around a lot, so whichever church was in town we went to. There is only one church so who cares which. That was very liberal minded for that time."

Barbara was impressed by the Catholic Church compared to the "austere" United Church in Wasing. "I thought the Catholic Church was a real cathedral, it was huge and very ornate. There were a lot of wood carvings from Quebec. I remember as a child thinking these statues were marvelous."

Getting by

There were advantages to living on the edge of civilization. In the summer, gardens provided cheap food, and the forest provided meat, in and out of season. Donat Huneault recalls that they "went hunting all the time,

there was lots of rabbits, and if we happened to kill a deer or something we'd bone it, steam it and put it in big quart jars with a little bit of salt." Many families had pigs and cows to provide meat and milk. Barbara's mother raised chickens and used to give some of them away.

But when winter came around pantries and root cellars soon went bare. "I remember my mom crying lots of times, not enough food, not enough money." recalls John Rouselle. "I mean, when you've got three or four kids, well it's hard. I think we got along pretty good. We had hand-me-downs and we used to wear rubber boots, summer and winter, and I was cold in winter."

Barbara Scott was one of the lucky ones. Her father had a job. She recalls her confusion as a child when she visited one of her friends near Christmas. Although they had a huge candle that had been "blessed by the Pope" in their house "they didn't have any money for Christmas or Christmas dinner so I went home and I said to my mom and dad, 'They haven't got any Christmas, can we have them?' So we had the whole family because they had nothing." As a child, Barbara couldn't understand why they had spent what she thought was so much money on a "stupid" candle. Now she says, "Looking back as an adult, I'm wondering if during the Depression that candle wasn't the whole thing. They had a lot of pride, you see, so the money all went into the candle, but they didn't have enough money anyway. I got the impression that everybody was pretty poor."

The families in Fossmill learned to expect little for Christmas during the Depression. "Christmas used to come around and we had an apple or a few peanuts in a stocking or something," recalls Roger Bergeron. "You were glad—as far as I was concerned anyway."

After Christmas, Barbara learned another lesson. "My aunt and uncle came up and brought us some oranges; in those days we only got oranges at Christmas. I took one to school and the kids fought over the peelings. I never took an orange to school again, I never took anything special that they didn't have. We weren't that well off either but we had a steady income, you know."

Later when she went to the Wasing school she discovered that the children of farmers in Chisholm were not any better off. She remembers that at Wasing "nobody brought lunches. They had a stove down in the basement, and everybody would bring something. We would throw it into a big pot and make sort of a big goulash for lunch. The teacher, Aileen O'Neil, used to bake bread for us."

Tent caterpillars and earthquakes

The Depression was not the only source of concern; nature joined in complicating peoples lives. "We had a terrible infestation of tent caterpillars," says Barbara Scott. "They were so bad, my father dug a trench around the house and put gasoline or kerosene in it and burned them. Our house was just covered with them—the whole town! A train went off the track, it was so slippery. The tent caterpillars were just unreal."

Max Smith remembers that a train was slipping so bad on the grade at "Cooper's Cut," near Wasing, that the CNR brought in a second locomotive. The first was on the front of the train where it crushed the caterpillars, while the train crew poured sand on the track. The second was pushing the train from behind.

In 1935, the area was hit by three earthquakes. They caused some damage in North Bay, but Fossmill was only shaken. Barbara remembers hearing "stuff rattling in the kitchen and dishes falling off the shelves."

Hoping for a future

The community of Fossmill struggled through the middle years of the Depression, concerned with survival, but also concerned with staying together and with welcoming new arrivals. Barbara's family was welcomed into community life. "I remember surprise parties they used to have. They'd come and everybody would bring food, and they'd bring some instruments. They'd take turns going to different peoples houses—like a Shivaree," she says, comparing the party to the traditional noisy celebrations in honour of newly married couples.

In Montreal, the Staniforths were weathering the depression somewhat more easily. Donald, who was only 13, recalls that his father "had to get a job down in the States, things were pretty close. We didn't suffer or anything, but we weren't living a luxurious life. You'd get a new pair of skis or something, and you wouldn't want to break them." He says that they didn't notice a drop in the standard of living, but he recalls that "It was worse out West. We had relatives who we sent clothes and stuff." He says Fossmill "was over. It was finished there, and Dad had to figure out what to do—he didn't have much money."

Sydney was interested in the hardwood trees on the old Booth limits inside Algonquin Park, and he was interested in Booth's logging depot beside the CNR on Lake Kioshkokowi at Kiosk. While working in the United States, he had Jack McGibbon cruise and evaluate the Kiosk site. Stella McGibbon recalled that Jack "was never off the job. Fossmill was done. They were looking for more timber, so Kiosk was the place, and that's were he went to investigate." Sydney offered Jack shares in a new company he was planning to start. Peggy remembers that Sydney raised capital from his contacts in the United States. Soon he would be ready to make an offer to the J. R. Booth estate.

The people of Fossmill listened to the rumours of a new operation with cautious enthusiasm. Perhaps the men would be working again and perhaps the dark cloud hanging over Fossmill would clear and life would take a positive turn again.

Waiting: *The children of Fossmill hoping for a future.* *[MG]*

William Mackey and J.R. Booth at Kiosk

Mackey Camboose: *Early lumberjacks lived in these low log buildings heated with a huge central fire pit.* [LMu]

The Amable du Fond River was the heart of the logging area of the Staniforth operation. Starting in the 1850s, William Mackey took white pine as squared timber down the Amable du Fond. To cut white pine into lumber a sawmill was built in Calvin Township east of the twin bridges between Smith and Moore Lakes. As a result of the mill, the village of Eau Claire developed there. Mackey's roads opened the area and his road along the river to Kiosk later became Highway 630. The construction of the CPR in 1881 provided economical transportation. Mackey's agent Andrew Ryan and his family ran the store and post office there for years after Mackey sold his holdings to J. R. Booth in 1902.

A May 15, 1885 *Canada Lumberman* article provided romantic insight into the Mackey operation. The author wrote about "the roll and dash of [the Amable du Fond's] rushing water striking steadily on the ear, intermingled with, at momentary intervals, the sharp hissing and cutting sounds of the circular saws as the logs are driven against them," and said, "It is a sawmill so complete, well-adapted, and thorough for its particular purpose, as to merit description. Here, above the water, stands a young habitant, bright in parti-coloured cap and jacket, his pike-pole guides a monster floating log near the spiked chain, it feels the point, it shivers, and seems feebly to resist its fate, as it is dragged up into the mill and deposited on the platform; rolled thence on a movable one, which bearing three men, passes rapidly back and forth beside a circular saw, it passes up and loses a slab, back again, another one is gone, and both run automatically along a platform of rollers outside.

Eau Clair: *This settlement on the Amable du Fond River in Calvin Township developed around the William Mackey operation.* [CT]

Eau Clair Gorge log chute: *Soft wood logs were floated to mills on rivers and lakes. In order to get past rapids and waterfalls, long ramps were built and the logs flushed down them to open water.* [CT]

The remaining portion is run to another platform, manned likewise, and cut into boards; they pass to a stationary one covered with rollers, where a small saw squares their ends, and at this point they are out of the mill and borne on tramways to the board piles. About 50 men are around the mill. It is the most busy of scenes. All day the logs climb in—all day the boards pass out."

J. R. Booth, "Canada's Greatest Lumberman" had extensive operations across Northern Ontario. He bought Mackey's limits and soon acquired further limits south of Kiosk in Algonquin Park. He took his logs down the Amble du Fond River to the CPR until the CNR went through the Park in 1915. He shipped from Kiosk as well as at Fossmill to the west of his limits. One report states that "The cut from Camp 2 was hauled to the railroad [CNR] and loaded on cars and shipped by rail to Ottawa. To carry out the loading operation, a steam Jammer was used and it was necessary to enlarge the switch line already built at Fosmills to accommodate at least 100 box cars." Booth used steam Jammers both at the skidways and at the siding to load and unload logs.

J. R. Booth's Kiosk Alligator: *These flat bottom tug boats towed logs across the lakes. [UTA]*

J. R. Booth's Kiosk office in 1922 *[UTA]*

J. R. Booth logging camp in 1922-23: *Camp near Kiosk prior to Staniforth's arrival. [UTA]*

Old J. R. Booth office in 1936: *Sidney Staniforth was interested in the abandoned depot and logging limits at Kiosk.* [DS]

11

NEW HOPE

"No telephones, no doctors, no police"

Kiosk was a deserted place in 1936. It was on the CNR inside the north boundary of Algonquin Provincial Park where Lake Kioshkokowi empties into the Amable du Fond River. The William Mackey and the J. R. Booth lumber companies had stripped the area of white pine for squared timber and saw logs. The abandoned J. R. Booth bunkhouse, store, office, and warehouse were all that remained.

"I was one of the first people they took to Kiosk. There was nothing there, only a barn and three buildings. There was no station, they moved that in afterward," recalls John Gleason. "I went about two weeks before anybody else came along. I went with Mr. Denis. We fixed up some of the windows and stuff." Soon more carpenters would arrive and Kiosk would be reborn.

Sidney Staniforth had an excellent relationship with the family of the late J. R. Booth. He had approached them to see if he could access the hardwood that remained on their

Boarding house and store: *only three J. R. Booth buildings were left on the Kiosk site.* [DS]

Circular-saw mill: *A small mill was built first to cut the timbers for the big bandsaw mill.* [GL]

Camp 2: *An early Staniforth Lumber Company logging camp.* [GM]

New mill: *The surviving Barnhart loader from Fossmill was brought in to help with construction.* [LN]

limits, and to arrange for the use of their Kiosk Depot. C. Jackson Booth, J. R. Booth's oldest son, became president of the Booth Lumber Company in 1925. Another son, Fred, and daughter, Gertrude, were directors.

"When Dad set the company up at Kiosk he didn't have much capital, but he went to the Booths and he got the cutting rights," recalls Donald. "He was able to cut but he didn't have to pay the stumpage 'til he practically sold the lumber, so it helped to finance the thing."

Once he had access to the limits, Sydney formed the Staniforth Lumber Company and brought in a crew, mostly from Fossmill, to build a circular saw mill, which cut timbers for the larger bandsaw mill he planned. The men left their families in Fossmill during the week and stayed in the bunkhouse that John Gleason and Albert Denis had prepared.

When the bandsaw mill was finished, the logging and the milling of lumber began again. Over 70 people from the Fossmill area were hired to work in the new mill and in the bush operation. At first the new company did not find it necessary to build houses in Kiosk, because houses, a school, and a church were available in Fossmill. Instead the men continued to stay in the bunkhouse and return home on weekends.

Depression wage levels were low. Roger Bergeron recalls, "When I started in Kiosk in '39 I was getting 17½¢ an hour and I worked ten hours a day, that was $1.75 per day. The men made $2.25 a day. I paid 75¢ board at the boarding house, so in my case I had a dollar left per day. So if the month had 25 days, it was easy to figure out my pay."

Home on Sundays

Roger recalls the 14 mile (23 km) commute back and forth between Fossmill and Kiosk. "My family was in Fossmill and I worked in Kiosk. We took the Transcontinental to Fossmill at four o'clock on Sunday morning and we spent the day there. It cost 50¢. They had a special car set aside just for us. Sometimes there were 60 or 70 people for Fossmill and Chiswick. We would go back on the Transcontinental at one o'clock on Monday morning."

Myrtle MacDonald remembered her brother-in-law, Lawrence Huneault, coming home to Fossmill from Kiosk. "We'd all be in bed and when we'd get up in the morning he was there," she said. "He'd spend Sunday and then he'd catch the 'Passenger' back to Kiosk. There were some that missed it and they'd catch 'the mixed' going through on Monday." Work finished at 6 p.m. on Saturday. In order to get home earlier, some men walked the 14 miles to Fossmill instead of waiting 10 hours for the train.

In the winter, many of the men working in the bush did not even come home on weekends. "My dad wasn't home that much," says Bernice Martin. "My mother was home alone a lot because Dad used to go to the lumber camps all winter long. He would leave in the fall and come back for Christmas, and then he would be gone from Christmas until the ice broke. We saw him, but it was not like other fathers."

Barbara Scott was an

outsider and recalls her impressions of that time. "I can remember there was a lot of drinking when they came home on the weekends. I think that was a very turbulent time for people—transition always is—no stability."

For most, the next ten years would be a difficult time in Fossmill.

The "Red House"

Amedee Laferriere was one of the workers who went to Kiosk. He was not happy leaving his wife Rachel and their children, Gaetan and Denise in Fossmill. In 1936, the CNR reorganized its track maintenance sections and abandoned their section house at Coristine, about a quarter mile west of Kiosk. Amedee purchased the house for $15, and moved Rachel and their children from Fossmill.

The "Red House": *Rachel Laferriere and children Denise, Robert, and Gilles on the porch of the first house in Kiosk. The wooden doll carriage on the right is her father's handiwork.* *[GL]*

"When we first went to Kiosk we used to live in the Section House. That was the only house there," Rachel said. The house was painted the deep red of the CNR boxcars so it was called the "Red House". Later Amedee's brother, Oscar, married Elise Desjardins from Fossmill, and joined Amedee in the "Red House". Later, Uncle Ovilla and his family lived there for a while.

There were only two women in Kiosk at first. They were "Mrs. Lachance, the cook, and I, that's all we had there," Rachel said.

Freight Train: *CNR train passes through Fossmill going east. The school is on the right. [MG]*

Fassett fades away

By 1936, the Fassett Lumber Corporation—now without Sydney Staniforth—went into voluntary liquidation and was all but dead. The lumber was sold and the last of the logs cut. Tom Howard collected rents, ran the store, and was selling off assets. In March 1937, the Fassett Lumber Corporation became the Fassett Lumber *Company* and the offices were transferred to Fossmill from Montreal.

School

Without the Fassett Lumber Corporation's taxes, the Public School Board closed the Fossmill school and over 30 students were left without a school. "Of course they were permitted to attend the school at Wasing, some two miles away, to which they would have to walk on the bush road," wrote the local Priest, Father Gravelle, in *Our Chisholm Story*. "Many of these small children were French speaking and could understand little English. When Inspector Brown was consulted by the teacher as to what she was to do with these pupils, she was told to just let them keep on attending classes until they learned English and could follow the classes."

Neither the farming families in Wasing nor the Fossmill residents were happy with this arrangement. Consequently, at a meeting of the public school ratepayers, Father Gravelle's proposal that the Fossmill school building be turned over to a new Separate School Board was unanimously accepted.

In September, 1937, the Fossmill school reopened as a bilingual separate school. "No money was on hand to open this new school house, furnish it and start to pay the wages of a teacher," wrote Father Gravelle. The secretary of the Board applied to the Department of Education for a special grant of $600, the board was promised $100 by Fassett, and a fee of $5 per family was imposed. This was enough for the school to continue.

Father Gravelle also arranged for a telephone line to be run from Fossmill to Chiswick. "It was just like a bush line swung on trees. It was in Aunt Maggie's house, so if you wanted to call out, you could go there," recalls one resident.

The year 1937 was full of significant events in Fossmill.

Allan Gleason finally made his dream come true. He retired and moved to a farm he bought in Chisholm. "He didn't want to go and work anywhere else," recalls his son John. "He was getting up in years then. He had that farm for a long time. He didn't do much with it, but he had it sitting there."

Allan's daughter, Margaret, married Chisholm's Stan Frederick and had a big reception in Fossmill. The spring wedding date

Separate school: *Students on the east side of the new Fossmill Separate school after Fassett has left. [BL]*

Students: *Fossmill children continue to have a place to go to school in Fossmill when the public school is turned into a separate school. [CWI]*

Gleason family: *Grace, Walter Clark (Nora's husband), Allan, Fran, John, Margaret (her son, Jack, is sitting on left), Dorothy, Nora, Lester, Hazel (Lester's wife), Harold, and niece Edna Clark with dog.* [MF]

Three generations: *Allan Gleason with daughter, Margaret, and her son Jack.* [MF]

Howard house: *Margaret Gleason and Stan Frederick moved into the former Howard house. A second story dormer put on by Howard was removed because of leaks.* [MF]

however, did cause one problem. One resident recalls that "There were more blackflies than people" at the wedding. Margaret and Stan moved into the former Howard house.

On May 12, long time Fassett employee Morris Donahue died quietly in his sleep. "He went up to bed after dinner, and the saying of the rosary—later, when he was called he didn't answer," recalled his granddaughter. Earlier that day he had helped Man Martin get Tom Howard's truck out of the mud by the school. "Morris and my dad were together the day he died," says Lionel Martin.

Alderdale disaster

On an early summer day in 1936, Maggie Donahue, who was the new postmaster, was standing outside her house in Fossmill talking to Louis Polichuk, the railway section foreman at Kiosk. Louis had come to Fossmill to get his mail. Just after 3 p.m., CNR freight train number 401 from Montreal passed Fossmill on its way to North Bay. It was made up of 32 cars, some carrying coal, some carrying bricks, others carrying beer and general merchandise, and one carrying gasoline.

Louis recalled later that 20 or 30 hobos were also riding on that train. One of the hobos was 49-year-old Andre Lamgron and a friend of his, both from Hawkesbury. Another hobo, who was from Montreal, had just removed his boots to make himself more comfortable for the long ride.

Minutes later, as Maggie and Louis talked, they heard a loud explosion coming from the direction of Alderdale Station. Soon they saw a thick cloud of black smoke spiralling into the sky.

"On June 27, 1937 at 3:30 in the afternoon, there occurred in Chisholm, the worst train wreck in the history of the 'North Country'," wrote Stanley Anderson in *Our Chisholm Story.* The first car of the train had derailed on an embankment just east of Alderdale, and the rest of the cars plowed into it, spilling "in a tangled mass of ruin on either side of the high embankment." The gasoline car exploded. "A massive pillar of smoke and flame towered skyward enveloping the whole scene and promising more destruction," wrote Stanley.

Local people rushed to fight the fire and to protect the barn of a local farmer, but instead

***Alderdale disaster:** A CNR freight train, carrying hobos, bursts into flames when it derails east of Alderdale station.* [NBN]

were confronted with human tragedy. Stanley wrote that "A transient who had been catapulted from one of the cars ran burned and screaming from the wreck. His clothes burned off, he suffered terribly and it is said he begged to be killed." That transient was Andre Lamgron. He was rushed to hospital and survived, but his travelling companion and another man were killed in the blaze.

Lloyd Graff, a local farmer, told the *Nugget* that he and his brother found a man from Montreal walking shoeless down the road. "My brother brought him into the house and we rigged him out with a pair of shoes."

"They were just a few of the hundreds of unnamed and unwanted young men who daily 'rode the rails' from city to city in the depression times looking for work," notes Stanley Anderson. "They were members of a 'lost generation' who would be called upon to make a further sacrifice when war broke out two years later."

Fossmill children: *Roger Bergeron (back) and the Rouselle children. [RB]*

Cleaning up

As Fossmill began its slow demise, and the Fossmill men began working in Kiosk, locals found work helping Tom Howard clean up the Fassett mill site, camps, and railway.

"I was only about ten when my dad and I went in and loaded the old boiler and different stuff," remembers Lloyd Tennant, who still lives a couple of miles from Fossmill on Wasing Road. "We also loaded stoker timber, hardwood tree lengths that they took to the mines and cut up for mine poles. There were quite a few families that lived there. They cut wood in the bush and I would go in with the team and draw it out to different houses. The track was pulled up in the early '40s. Some men went in and tore the old camps down and brought them out before they lifted the rails."

"When Fossmill went out of business they hired us to bring all the stuff out from the camps," remembers Max Smith, now 80, who still lives on his farm a mile from Fossmill. "In 1938 we cleared it all out and loaded it on the flatcars and brought it out. I used to go in with Tom Howard." Max remembers that to get in to the camps on the railway tracks, Tom Howard mounted the front end of an old 1929 Plymouth car onto a railway lorry, and connected the engine to the lorry's flanged wheels. On the back he built an open wooden box for the workers and their tools. "They had a problem turning it around," Max says, "so they got the bright idea to put a jack in the centre of the car. Then they jacked the car up and turned the thing around and home we would go." He also says, "We had to take horses in to draw the timber out. We had to walk them from Fossmill for three-and-a-half hours. We took out equipment, all the blacksmith's stuff, anything that was any good."

"Scrap dealers bought the locomotives, because I checked into that when I was working down in Toronto," remembers John Gleason. No. 54 was eventually sold to Zagerman's, a railway scrap dealer, and was for sale in Barry's Bay in 1942. Railway historian Colin Churcher wrote in a May 1988 article[17] in *Branchline,* that "Details on its final disposition are unclear, it was probably scrapped as part of the war effort."

Making the best of it

In the years before the Second World War, the women and children in Fossmill tried to make the best of the situation. "The woman used to meet and chit-chat. I think they were all pretty busy. My mom was a very busy lady—well she had six kids and no conveniences," recalls Bernice Martin.

"After the mill burned down the general store stayed there for a little while, but not very long. There was nothing in there. Only a little store that's all—milk, eggs, bread, and stuff like that," recalls one villager. "People used to leave Fossmill and go down to

Charboneau's general store in Chiswick. When a few started getting cars, they'd go to Alderdale and to Butler's store."

Some people walked to Alderdale and returned on the local train with their load of groceries. "The mixed" became the centre of activity in the village. "It was only the mixed we met. That was big doings, that was," recalls Myrtle MacDonald (Coulas). "We'd all go and wait for the mixed to come. Our groceries would come and we'd see who was getting off—not too many people ever got off it—but it was the thing. Our orders from Simpson's came on the mix."

Food was also kept on the table by hunting and fishing—sometimes out of season. "What I remember when I lived in Fossmill was that we lived on venison most of the year," recalls Margaret Frederick. "My husband's brother had a farm at Wasing and around the first of July the deer used to come into his grain fields, and he would get venison. He would divide it all up. We all got some and we'd can it. Sure they had hunting seasons. These guys are dead now, so they can't do anything to them. My husband lived on fish all summer in his lunch. Every evening he would go fishing down at the Wasi Creek and he would come back with trout, beautiful dark red, and he would say, 'Fry a couple of these up and put them in my lunch tomorrow'. We didn't have any electricity, and even though we had ice it still didn't keep your meat."

The radio provided a link with the outside world, as well as some humourous memories. "We were the only people in town that had a

Good drainage: *Marcel Levis, Gaetan Laferriere and friend pose beside an outhouse looking over the Wasi River.* [GL]

"The mix": *Local train picked up passengers and local freight on its journey between Brent and Capreol.* [WL]

Roger and Laurette Bergeron: *Roger returned to Fossmill each weekend. Soon he would build a house in Kiosk.* [RB]

radio for the longest time. People used to gather at our house," remembers Bernice Martin. Her brother Lionel recalls, "People started buying radios, 'Marconis'. They were three feet high. To get better reception we would take an aerial at night and throw it over the telephone wires that went into the Donahue's house. My grandmother phoned the Donahues to talk to my mother some evenings. After they talked my mother would come off the phone and she'd say 'there was radios on the telephone'—she couldn't figure out why!"

Ila Huneault (Legacé) says, "Mrs. Suave, my mother, Mrs. Bergeron, and my aunt, who had the radio, would congregate and listen to the soap operas." Batteries for the radios were expensive. One of the women's husbands "would make her mark all the hours down, so she would always go to other women's houses to listen to the radio. They'd see her coming and the women would get mad. They'd gossip. They'd say, 'look at that, she comes over here to listen to her stories.'"

Hard candies and Santa Claus

"Christmas was great," recalls Bernice Martin. "We all had to do something in the Christmas concert. Maggie Donahue was the only one who had a piano, so we went to her house to practice the dances and songs. Then, on the last day of school, we performed in the Christmas concert. The teacher was behind the curtain coaching us, because by the time we got up there we forgot our lines. We sang 'O Canada' and all the rest of it, you know. And then Christmas would come together and it was like magic to us. All the shopping was done out of the catalogue and the parcels came to the Post Office at Maggie Donahue's who used to hide the parcels. We went for the mail, thinking we were going to catch her. Come Christmas day there would be hard candies by the bucket, and ice cream and lots of food. The snow was so deep and so white! Everything was so beautiful. I can remember one Christmas, Stanley Frederick dressing as Santa Claus and going to all the houses. That just blew all us kids away."

Cut finger and Dr. Dillane

Throughout the history of Fossmill, getting to the doctor in Powassan was always a problem. John Rouselle told a story full of bad luck. When he was about 12 he was helping his father cut wood in the bush when he cut his finger with an axe. His father sent him home, but when he arrived home he discovered the end of his finger had been cut off.

When his father came back looking for him he said, "Christ, you have to see Dr. Dillane" in Powassan. They borrowed a car, but just past Chiswick the car stopped and wouldn't start. "A guy came by and said he knew what was wrong and opened the hood and took out the timer and cleaned it." John said. "When he put it back the car started right away and we headed for town."

When the doctor discovered he had cut the

tip of his finger off, John said Dr. Dillane wanted to cut the finger off at the joint. "I told him it was short enough." John says. The doctor said, "You little son-of-a-bitch." John said, "That's the way he talked. 'You're going to be back in a couple of weeks.' So I went to school and I used to hit myself and it used to bleed and it was sore and the bone was sticking out. I said there is no way, I got to go back. When I went back, Dr. Dillane smiled and said, 'I told you so you little son-of-a-bitch.'"

"Just like hillbillies"

When Elva MacDonald married Lawrence Huneault they built a house east of the village. They found a spot, between the fire rangers' cabin and a swamp, where there was a spring and soil for a garden. Unlike the rest of the houses, which were built out of cull lumber from the mill, they built their house out of logs. It had two bedrooms in the back and a front room that served as a kitchen, dinning room, and living room. They covered the walls with cardboard, papered over that with pages from old catalogues, and then covered over that with wallpaper. In the winter they banked sod around the base of the house to keep out drafts.

"My father used to trap," remembers their daughter, Ila. "He never worked full-time in the mill. Later he worked in Kiosk in the summertime and trapped in the wintertime." During the '30s and '40s workers made $30 a month in the logging camps and Lawrence got $25 for each beaver pelt. As a result Lawrence went into the bush to trap for a week at a time. At night he slept in the open "between two fires."

Because they lived so close to Algonquin Park, the temptation to trap and hunt outside the law was powerful. During the Depression there was an increase in poaching for furs in Algonquin Park. "I remember my dad saying to my mother, 'El, I'm going to be coming out such and such a day,'" recalls Ila. "He said, 'If the park rangers are around town, shoot a couple of shots at such and such a time and I won't come out of the bush.'"

"They came and tried to find his furs," recalled Myrtle, Lawrence's young sister-in-law. "We took the mattress off the bed and lined it with mink, martin, beaver and everything. They would say, 'Pretend you're sleeping'. The game warden came in and looked in the bedroom and said, 'Oh, the kids are sleeping. I'm not going to disturb them.'"

Myrtle also remembered that a pet goat saved the day on one occasion. She said, "We had a goat and it was a real pet. Anyway, the goat would lay on the verandah every night. When Lawrence came home on the weekend he would go outside at night and fall over it. He finally said, 'That goat is going to die.' So one Sunday morning we were all ready to eat breakfast and he says, 'Would anybody like some goat meat?'—Well, we knew our goat was dead. Nobody would eat it, so he decided to go out and hunt a deer. In the meantime, we skinned the goat and put the hide on the side of the log shack. Lawrence got a deer and we were boning the venison, when suddenly the game warden came around the corner. Here we had this great big washtub full of

"On the brow of the hill": *John Martin in the early 40s in front of one of the houses on the hill across from the old lumberyard. The houses would soon be torn down for the lumber.[BL]*

venison. He looked around and he says 'Oh, I see you killed your goat' and Lawrence said, 'Oh, yeah.'"

"My mum was a good shot," Ila remembers, "When my mum shot a deer Mr. Rouselle would hear the shot and he would run over because he knew he would get some meat."

In 1936, elk were released into Algonquin Park, where they apparently survived for only a few years. One reason may be obvious from the following story. "Mom shot an elk once, just outside the house. We had a big swamp at the bottom," Ila says. "Mom saw an animal

***Elva Huneault (MacDonald)** [MC]*

and kept putting her gun down and then sighting it. She said, 'Is that a cow or not,' and, 'I don't think it's a cow.' So she finally let him have it. She cut the head off and we kept it in the creek so she could show Dad, when he got home on the weekend. We didn't tell the game wardens."

Ila tells two other stories about her family. In the first she recalls the time her father returned after hunting for deer. "I remember him coming out and saying to my mom, 'El we're going to go to the salt lick.' He'd put salt down, to get a deer." After the deer was shot the family had to butcher it quickly. Ila says, "I'd be holding the flashlight and they'd be skinning the deer. It was dark and because there was nobody at home they'd say, 'If you see the door opening at home let us know because we'll put the flashlight out and stop working.' That was survival."

In the second story, she recalls the level of fear the rangers instilled in them. "One time we caught a deer and we cut it all up. We put the steaks in a barrel and put grease over them so they would keep. My aunt had a big family too, and her daughters were over helping us. Then somebody said, 'The park ranger is coming.' My cousin, who was about 14, picked that barrel up and put it in the bedroom—the adrenaline rush—it took three of us to bring it back out."

"We were just like hillbillies," is how Ila sums up her time living on the edge of civilization. "It's something I'll never forget though."

Curiously there were some people who tried to live even farther back in the bush than

***Fossmill Ranger cabin:** Eileen, Lillian, and Ila Huneault in front of the cabin overlooking the forest south of Fossmill. [IL]*

***Hunter and trapper:** Lawrence Huneault on the right and Leo Rouselle show off their catch. Lawrence supplemented his income from the mill and bush with income from trapping and hunting. [IL]*

the Huneaults. Perhaps they were running from something in their past or searching for some peace. Ila remembers one woman who lived "way back in the bush by herself with her daughter. She got her bag of flour and her stuff from Alderdale, and they'd drop it off at Fossmill. My mum said, 'We can't let her go down there, she can't carry a bag of flour way back there.' We carried all her sugar and whatever, and my uncle carried the bag of flour—she would have done that all by herself!"

Roger Bergeron remembers an earlier set of refugees. "A family with three or four kids lived in an old house in the bush a mile from Fossmill. They never worked at the mill. It was unbelievable—the poverty."

Isolation

While the men were away working in Kiosk, the women and children in Fossmill were vulnerable to loneliness, isolation, or worse. Bernice Martin recalls the isolation. "No telephones, no doctors, no police, and the women were there on their own with their children."

There were a few cases when men tried to take advantage of the women's isolation. Ila Huneault recalls that one man thought he could take advantage of her mother. "This guy propositioned my mother. He said, 'I got a fisher [a valuable fur] which you can have if you give me something.' My mom sent him away." Ila recalls one time another man made the mistake of underestimating her mother. "My mom and us kids were working in the garden one day when we saw someone at the ranger cabin. Mom says, 'He's not doing what I think he's doing, is he?' She grabbed her gun and yelled, 'Put that thing back in your pants and get out of here,' and she fired a shot over his head. A neighbour down the road commented later, 'I don't know why that guy drove past so fast. He nearly killed some of my chickens!'"

"My brother, Irving, was in Powassan at the hotel," another family member added. "This guy came flying in, in his car and went into the hotel and said, 'Some crazy lady in Fossmill took a shot at me. She could have killed me,' and my brother, realizing who he was talking about, said, 'If she wanted to kill you, you'd be dead now.'"

One feisty Fossmill woman told how some of the women dealt with a married man from Kilrush who came visiting

Women and children: *With the men away in Kiosk the women and children were left alone in Fossmill. Sixteen-year-old Bernice Martin with her sister in 1943. [BL]*

another woman in Fossmill on his **Velocipede**. The women visited a nearby outhouse and left a message smeared all over his machine!

"I can remember when they thought anybody from Fossmill was bad. 'Oh! that girl from Fossmill'—the farmers in Chisholm they figured everybody—I don't know, I suppose we were all 'Hs'—as far a they were concerned—we were 'bad girls'," recalls one villager. "They couldn't even realize that we came from good people, that we were good people—it was Fossmill and we were wild."

Young women: *Group of Fossmill girls in 1942.* *[BL]*

Off to war

Soon the hardships of the Depression were replaced with the hardships of war. When the Second World War began many of the young men from Fossmill and Kiosk enlisted. Donat Huneault recalls, "The Dwyer boys, the Danis boys, and my friend Tony Hawk, they all went. I remember a bunch coming out of the bush once saying, 'Hey, we are going to enlist, we are going to war.' They would have had to go anyway. They were all young guys. They didn't have a choice."

The lumber company got exemptions for some of the skilled men because the industry was seen as necessary for the war effort. "It just depended on what job you had if the company would vote for you," recalls Donat, who was exempted for medical reasons. "I went to the doctor. He give me a paper. He said the 'only place you could get stopped is when you get on or off the train. All you have to do is show them the letter and they will let you go.' I never was bothered."

Donat recalls that some of his friends didn't come back from the war. One of them was Ambrose "Mike" Dwyer. He also says that some of the ones who did come back were in "pretty bad shape."

The home front

At home in Canada, Sydney Staniforth went to Ottawa and volunteered as a "dollar-a-year man" to head The Veneer Log Supply, a crown corporation that supplied veneer logs for the war effort, including the birch veneer used to build the Mosquito bomber. He was awarded the Order of the British Empire for this work (see portrait of Sydney with the medal, bottom-centre page 200).

During the war, the mineral Mica was in demand. Mica divides into thin translucent fire proof layers and was used for insulation and radio tubes. There was an open pit mica mine in the hills behind Fossmill. "We had a prospector looking for mica living with us," remembers Ila Huneault. "I remember going back and picking it."

Myrtle MacDonald, who was a teenager during the war, recalled a party with some soldiers in the abandoned Fossmill church. The soldiers were on duty on the CNR at

Going to war: *Silvio Danis poses in his new uniform before going to fight in Europe.* *[RD]*

Kilrush "protecting" the bridge there. In their spare time they used to visit Fossmill. "The church was tilted over and ready to fall down," Myrtle recalled. "Some of us and some of the soldiers broke in one Sunday and were acting up something terrible—we were pretending we were getting married! The next day the priest came in to drive the teacher to school. He got wind of us being in the church, so he came to the school and said, 'Since you are so smart, you can clean out the church.' We had to go in on Monday evening and take out all the statues and candles and stuff. Nobody had used it in years. The church was all cracked off to one side and dangerous, propped up with planks."

Worse for wear: *Two villagers play in the bell tower of the unused Fossmill Church.* [LN]

Myrtle recalled that during this time the new separate school was used as the church. "They had an altar in the corner and a statue with candles around it. Father Gravelle would come in and they would put a sheet up in the corner and everyone would go to confession there, and then he would say Mass. Every night after supper, prayers were held at the school. We were the only Protestants there for a while. We used to go because it was a night out."

As a separate school, the curriculum also included religious instruction. Elva Huneault was a Pentecostal, so she decided to send her children to the public school in Wasing. Her daughter, Ila, recalls that some people in Wasing "weren't too keen on this because Dad, with his trap line, had kind of got in, you know, a little trouble with the trap grounds. They were on the board and said, 'you can't come.' So Mom says, 'We'll see about that.' Anyway, we got into the school—my uncle and aunt, my sister and myself, that was my first year of school. My mom used to come to meet us. When she'd hear us coming down the road, she'd run and hide in the bush. We'd say, 'momma must be around' because we'd see the pig and the cow and then the goat—they'd follow her wherever she went."

They eventually went back to the Fossmill school when Alma Chayer, the Fossmill teacher, relented. Ila reported that she said, "I can't stand you kids going by in the snow and the sleet. I'll take you back." She allowed them to attend without having to take religious instruction.

McGibbon Family

After cruising the Kiosk limits for Sydney Staniforth, Jack McGibbon was given shares in the Staniforth Lumber Company and was woods manager in Kiosk for many years. He walked the trail from Kiosk to Highway 17 and helped with a petition to the provincial government for what eventually became highway 630 and provided road access for Kiosk.

The McGibbon family moved to Powassan where Jack joined with his daughter Peggy's husband Joe Brushey to form the Stanley Lake Lumber Company, which had a retail outlet in Powassan for many years. Peggy became a teacher, and in 1999 her son Peter is serving his fifth consecutive term as the Mayor of Powassan.

Last days: *Eva Pitz in front of the McGibbon house in the 1940s.* [MG]

McGibbon house: *Mrs. Margaret Polichuk was the last resident of the original woods manager's house.* [MG]

Fifty years later: *What remains today of the McGibbon house. The pipe is the luxury indoor plumbing drain.* [PM]

One o'clock in the morning

Bernice Martin tells the story of one cold winter night when a newcomer arrived in the community. "My aunt, Alma Chayer, made arrangements with a student teacher from the normal school to come to do practice teaching in our school after the Christmas break. My aunt lived in Chiswick and the teacher must have got the days mixed up, so she got off the train at midnight on Saturday, a day sooner than she should have. We used to stay up Saturday night with the Huneault kids to listen to the Grand Ol' Opry, so we didn't go to bed until maybe one o'clock in the morning. This teacher gets off the train—there is no one there—there is no one in the railway station. The houses are all dark and she had never been there before. She pushed her suitcase to the only light, which was our house, and she knocked on the door. There she was standing there, the snow was up to her ears and she had this suitcase and every house in town is in darkness. My aunt was living in the Donahue house nearby, so we went over and lit the stove and got her settled in."

Kiosk grows

Kiosk grew slowly in the early years. Alton Douglas, the Manager of the plant, built the first house there and three families lived in the "Red House" at Coristine. By 1939 there were a few houses and 44 permanent residents in Kiosk. Eleven of them were school age children, so a school was built.

"When I went there in '45," says Mike Larochelle, who was hired as the Manager in Kiosk after the war, "there was a bunkhouse that held over 100 men. They came from Chiswick and all around." He recalls that the company was eventually compelled to build some houses for the men. "We had to build some, they wanted to get their wives there. While I was there, the company built around 40 houses. Some of them weren't winterized. They were for people who only came with their wives in the summertime."

"As houses were built, people moved in," recalls Roger Bergeron. He says some of them also built their own houses. "In '46 I was working in Kiosk and my wife was living in Fossmill, and none of us liked this arrangement." Roger and his father salvaged panels from old pre-fabricated camp buildings and built a house big enough for their families

Last villagers: *Leo Rouselle (left) with Albert Bergeron during Fossmill's last days.* [RG]

Leaving home: *Alderic and Eva Bergeron in front of the Levis' house in Fossmill one of the few with a flat roof.* [RB]

First Kiosk home: *Children in front of the house that the Bergerons built in Kiosk.* [RB]

Main Street Fossmill today: *Abandoned CNR tracks to the left. Nothing remains of the Fossmill houses on the right but depressions in the ground.* [DM]

Fossmill declines

Back in Fossmill, John Rouselle got the job picking up the mail in Chiswick. "I used to walk there at first. After that, the catalogue came out with a bike in it. My dad bought it for me. It was just like a $60,000 car to me," he says. "I brought the mail to Mrs. Frederick, she was the new postmaster. The government used to pay us $10 a month."

In 1946, the Fassett Lumber Company sold the main 96 acre Fossmill site to Tom Howard for $400, ending Fassett's connection with Fossmill. Allan Gleason died the same year, and daughter Margaret and family

Nature returns: *Hot pond and location of the Fassett Lumber Corporation mill today.* [DM]

moved to his farm. Elva and Lawrence moved to Kiosk in 1948. Lawrence worked in the mill and Elva raised the children. In the summer, she ran a day camp for the Kiosk children. They continued their interest in hunting and trapping throughout their lives. Their son-in-law Maurice Legacé and their grandson Kent still trap in Boulter Township.

As more and more houses became vacant in Fossmill some were torn down and the lumber was used to build houses in Chisholm and Powassan. In the end, the Rouselles got the contract to tear the last of them down. John remembers, "There weren't that many houses left at the end. I don't know, I would say 10 or 12. There was a guy from Montreal that bought all the old houses for scrap. My dad and I took all the houses apart. I think we got $50 a house. We took a house down, piled it all, lumber, windows and nails, and he sent a boxcar in to take it away to Montreal for sale."

Lois Donahue remembered that "Leo Rouselle, John's father, built a great big house on Fisher Street in North Bay from tearing down the houses in Fossmill. It was a great big two-story house. They were the last people in Fossmill."

Today the property is almost completely overgrown with only a few root cellar indentations in the ground and some traces of roads. The CNR line was abandoned in 1995 and the rails and ties removed. The hot pond is clearly visible, but is overgrown, and is home to Canada geese and beavers.

Barbara Scott returned to Fossmill in 1998. She says, "When I walked down the tracks it was as if there were ghosts from the past." Ironically, a current road map, which includes all the new road names in Chisholm Township, still includes Fossmill, even though it has been gone for 50 years.

Fossmill 1996: *The CNR line, soon to be removed, is in the foreground. The hot pond and Wasi River can still be clearly seen.* [PM]

Kiosk school: *Children in front of the first Kiosk school.* [LG]

Kiosk Catholic Church: *The bell for the new church was salvaged from the Fossmill Church* **[NBN]**

Kiosk houses: *Eventually the town had 80 houses and was the largest community in Algonquin Park.* [ML]

EPILOGUE

NEW HOME, AGAIN!

"The forest seems to be waiting until the villagers leave."

Although Fossmill itself was gone, the people of Fossmill continued to flourish in their new homes at Kiosk. By 1950 there was a permanent population of 186 (92 children and 94 adults) as well as 60 workers living in the boarding house. A Catholic church and a recreation facility were built. A town power plant was established on the Amable du Fond River nearby, and water was piped into the community. The town eventually had as many as 80 houses and 600 residents and is the largest community ever to have existed in Algonquin Park.

The logging and lumbering operation at Kiosk was larger and was operated differently than at Fossmill. Because trucks and tractors had improved in design and were now in wide use, a logging railway was not needed. The mill was built beside the lake so that softwood logs could be floated to the mill. "It was a pretty big mill, with a double cutting bandsaw. They produced 40 million board feet per year," recalls Donald Staniforth.

Sustained yield

After the war, the Ontario Government passed legislation requiring logging companies to cut on a **sustained yield** basis. The number of trees cut each year had to be balanced with the forest's ability to regenerate. This amount was called the **Annual Allowable Cut**.

"We were operating on a sustained yield basis in Kiosk," recalls Donald. "We were coming back to some of the areas that had been logged from Fossmill. We figured on about a 40-year cycle. The diameter of the trees cut had to be 16 inches or larger. You wouldn't cut anything under that. You had to put it in your annual cutting reports."

The Staniforth Lumber Company eventually acquired the Booth limits outright and the company later made a deal

Kiosk: *Expanded train station (centre) and the restored J.R. Booth buildings in background.* [DS]

with Gillies Bros. Ltd. for the limits to the west of the Staniforth limits (the same ones Fassett had lost in 1934) in exchange for Staniforth shares.

In 1941, the company built its first kiln to dry its lumber faster. Eventually an expanded sawmill and planing mill, more kilns, and a veneer operation were added. In 1949, the first women mill workers were hired to work in the new veneer mill.

Sidney Staniforth managed the company from his office in Montreal, but made regular trips to Kiosk on the company plane. He was known for his desire to be close to the operation and to talk to his workers. His sons Harold, Donald, and Robert were also active in the management of the company.

Unions

Unions were being established in many industries in the late 1940s, and the lumber industry was no exception. Workers Louis Demers, Donat Huneault, and others carried out a certification program with the International Woodworkers of America at Kiosk in the early 1950s. There were two strikes, one in 1950 and another in 1955. The union negotiated better wages, hours, and working conditions. One worker recalls that his wages went from 72¢ an hour in 1950 to around 92¢ after the strikes.

"They were able to cut the work week, and in

Brakeman to truck driver: *Man Martin was no longer needed as brakeman and became a driver for the Staniforth Lumber Company.* [BL]

Trucks: *As they became more efficient they replaced the railway.* [DF]

Donat Huneault: *operates the planer in Kiosk.* [NBN]

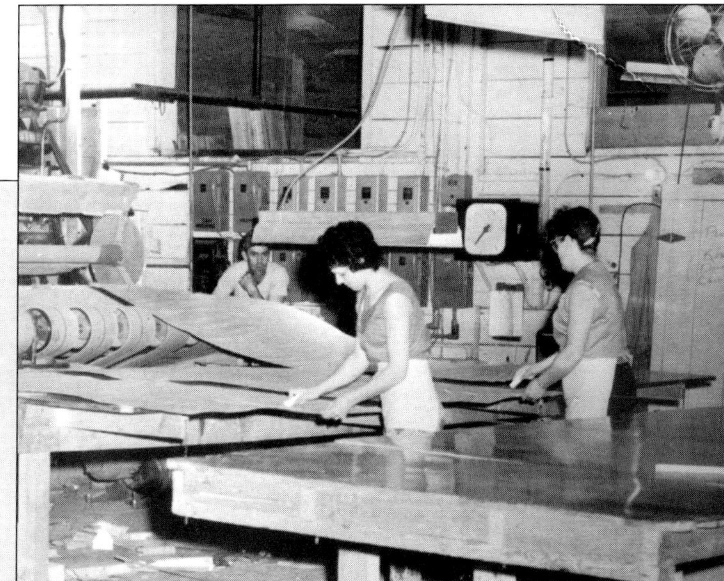

Women enter the workforce: *Women were hired to work in the new veneer mill. The woman in the centre is Vivian Laferriere.* [NBN]

Kiosk mill: *A view of the distinctive triple smokestack, jackladder, and log booms in the lake. The Boarding house is in the background on the right.* [GL/ML]

***"Log Pond":** At Kiosk, the Staniforth Lumber Company pulled logs up a jackladder from Lake Kioshkokowi. An area in front of the mill was separated with booms. In the winter steam from the mill kept this area free of ice, creating a "hot pond".* [APA # 1077]

those days a cent was a hell of a lot," says Donald Staniforth. "We used to fight over a half a cent. There was no inflation in the '50s at all, the cost of living wasn't going up and a dollar bought quite a bit in those days. We had to negotiate and then they had the grievances and stuff, and they kept wanting to take away our management powers."

Donald's father would not negotiate with the union. Instead he sent his oldest son, Harold, and a lawyer to meet with them. Donald remembers that his father "was a bit hurt. He thought he was treating them pretty well. That's what hurts—when you're running things and it's your own company. It's hard to take."

Company ownership

Sidney died as a result of a car accident in 1958, at the age of 73. He had provided 50 years of leadership through the Fassett, Fossmill, and Kiosk operations. He also

***Loader and tractor:** The surviving loader and tractors from Fossmill were put to use in Kiosk.* [LG]

provided leadership outside the company as president of the Canadian Lumbermen's Association, the Hardwood Lumbermen's Association and, during the Second World War, the Veneer Log Supply.

His son, Harold, became president of the company, Donald ran the veneer operation at Kiosk, and the youngest son, Robert, was in charge of the sawmill and sales. Alton Douglas, the first manager of Kiosk, was followed by Mike Larochelle and later Otto Koch.

In 1971, for a variety of reasons, the owners of the Staniforth Lumber Company sold it to the U.S. firm, Universal Oil Products. The three brothers remained as advisers for five years while they pursued other interests in the lumber business.

In 1973 the fate of the Goodman-Staniforth Company, as it was then called, was decided like so many other lumber companies—by fire.

Friday the 13th

On the evening of Friday, July 13, 1973, things were not going well at the Goodman-Staniforth mill. There were mechanical problems, and the 30 men on shift were sent home early. Ironically, a retirement party was being held for Amedee Laferriere, who had worked for the Fassett/Staniforth companies for 47 years.

At 11:30 p.m. Doug Farrow, the boiler room engineer, and James Callahan, the night watchman, were the only two men working. "I heard a swoosh and smoke poured through the doors. It came so fast I couldn't do anything," recalled Mr. Farrow. He managed to blow the steam whistle alarm and jam it open before he fled the building. "I pumped my legs, I'll tell you," he said.

The fire apparently started in a dryer in the kiln area. Plant maintenance superintendent, Don Falkingham, who doubled as fire chief for Kiosk, was at home when he got the call about the fire. As he held the phone it went

On strike: *Staniforth Lumber Company workers in 1955.* [LD]

dead, knocked out by the fire. Later he said, "We had no phone, no water, no power. What the hell could we do?"

Families, fearing their houses might burn, bundled themselves into cars, ready for a quick retreat. A string of empty metal boxcars parked on a siding between the fire and the town glowed red from the heat. If it had not been for the boxcars the flames might have burned some of the houses.

The mill's office manager, Jean Paul Lamarre, managed to call North Bay before the telephone lines went completely out. The North Bay Fire Department responded immediately with Fire Chief Art Haley, a pumper truck, and three men.

Realizing that little could be done to save the mill, everyone available concentrated on saving the hardwood veneer logs and heavy equipment. Alan Renaud managed to save the company tug and runabout by slipping their mooring lines and setting them adrift on the lake.

The fire burned 10 acres (4 h) of buildings and yards in 90 minutes. Fortunately no one was killed. In total, 234 employees, including woods crews, were put out of work, affecting 600 people when their families were included.

There was talk of rebuilding at Kiosk or nearby, outside the park, but it soon became evident that this was not going to happen. The 1974 Master Plan for Algonquin Park curtailed any mill construction in the Park. Park timber licenses were cancelled. Hardwood logging was to be carried out on a **selection cutting** basis by the Algonquin

Forestry Authority, a Crown Corporation.

Some people went to work for the Booth Lumber Company at Tee Lake in Témiscaming, Quebec, where the three Staniforth brothers Harold, Donald, and Robert were president, managing director, and vice-president respectively.

Other people got work at Rutherglen in the Universal Oil Products veneer plant, which is now Columbia Forest Products. Interestingly, today, Rachel and Amedee Laferriere's son and grandson both work at Columbia providing four generations of continuity to a story that began in Fassett, Quebec almost a century before.

Since Kiosk was a company town it did not have an elected municipal council. It did, however, have a community association which did most of the things a council would do. This association fought hard for the survival of the town, working with federal and provincial politicians, but to no avail.

The provincial government eventually bought all of the houses except four from the company, and gave the people until December 31, 1996 to move. Those who left received some compensation. As people left, bulldozers moved in and leveled the houses.

In 1983, reporter Gord McCullogh, wrote an article about Kiosk in the *North Bay Nugget*, indicating that only 25 of the original 80 houses remained. The church, store, and community centre were gone. The school was being used as the community centre. McCullogh began his article by stating that "The forest seems to be waiting until the villagers leave, so it can reclaim its own again."

Long before the December 31, 1996 deadline everyone was gone. An Algonquin Park Access Point office and the old park ranger's house are all that remain, other than a few foundations and tracks in the grass. Kiosk is now a provincial camp site and the ranger's house can be rented by campers.

Twenty-five years after the fire a Kiosk Reunion Committee still organizes an annual reunion. Former Kiosk residents and their families come from all over North America to play golf, have a fine meal, and reminisce about the incredible Fassett/Fossmill/Kiosk story and how they overcame hardships and difficulties to build, maintain, and hold together a community in twentieth century Northern Ontario.

Barnhart Log Loader: *Loader crew pose for the camera in this spectacular photo from Fassett, Quebec.* [ONA]

Appendix A: Equipment of the Fossmill Railway

One of the reasons we chose to study Fossmill and the Fassett Lumber Corporation was that it included a logging railway. Fossmill's location at the end of a country road at the base of the Northwest Algonquin Park Highlands included rough terrain and harsh weather conditions. Because the railway was a small short-line railway it also included interesting railway equipment and operations.

When we began researching, little was known about the locomotives and equipment. Local history books did not include the usual photographs of proud crews standing beside their locomotives, log loaders, or log cars. What types of equipment did they use? What was their make and size, and how did they work?

Many former residents remembered that much of the equipment was specially built for the logging industry. When Peggy Brushey described the machine used to load the logs onto the log cars she said it was "not a crane really, it was called a 'log loader'. It had a boom on it. It was a unit by itself that turned on a pad." Local historian, Stanley Anderson, said that the locomotives werc "Geared a way down. They were very slow but they were very powerful." Max Smith remembered a trip on one, and in particular, the sound they made. "They were them old type engines that had the drive on the side of them. You could hear that old thing coming for miles and miles." Were they describing a Heisler, Climax or Shay locomotive (the three main locomotive types used in the industry)?

In one of our early interviews, 98 year old Stella McGibbon described the locomotives. "Instead of the big drive wheels and the big shaft like the railroad engines had, there wasn't any of that at all. It was a series of—I don't know, I'm not an engineer—a series of gears going up and down and they worked around. They were much smaller than a railroad engine, but strong enough to haul a good long load of logs."

This is a good layman's description of a Shay-

The left side: *This photograph of No. 54 is unusual because, most of the time, photos of Shays show the right side of the locomotive, which includes the cylinders and the gears. The long hose under the boiler is for filling the water tank from streams and lakes along the side of the railway .* [LN]

geared locomotive (see below for detailed description). But how big were they? The two locomotives used were referred to as "the little engine" or "the big engine". We did know that by the time the locomotives were used in Fossmill they were already getting old. In 1930, the Algonquin Park superintendent referred to them as "having seen better days."

After consultation with logging railway experts, we knew that we were looking for Shay-geared locomotives and Barnhart log loaders. We were also informed that there was a hard-to-find book by Michael Koch, called *The Shay Locomotive* which listed all the locomotives ever built by the Lima Locomotive Works, builders of the Shay locomotive. The book included the builder's numbers, dates of construction, gauges, cylinder and wheel sizes, and the name of the company for which the locomotives were originally built. The following entry appeared for 1910:

> 2353 10-18-10 STD 12 X 15-36. Fassett Lumber Co. No 54- Fassett, Que & Foss Mill, Ont.

This was the "big engine" used in Fossmill. There were also listings for No. 53 (used in Quebec) built in 1908, and for No. 51, the "little engine" in Fossmill, built in 1905.

Most of the railway equipment was brought to Fossmill from Fassett, Quebec, in 1924. Lois Nault recalled that her father Joe Donahue told stories about how her grandfather Morris Donahue, the railway superintendent, brought the locomotives from Fassett, Quebec. "Joe told me the story about his dad taking a locomotive and a bunch of cars and driving on the CN all the way out to Fossmill from Fassett and going back on the regular train and bringing back some more." Some of the equipment used in Quebec was sold to the Standard Chemical Company and never made it to Fossmill.

Locomotives

The Shay-geared locomotive was designed by Ephraim Shay in 1880 specially for logging and mining railways. They were produced by the Lima Locomotive works in Lima, Ohio from 1882 to 1945. A catalogue from the early 1900s states: "The Shay Locomotive was designed to operate on wood or iron rails, on heavy grades and sharp curves, and

Locomotive No. 54's bell: *Don Humphrey, relative of Lester Gleason, holds the bell salvaged from the No. 54 before it was sold. [DM]*

Special qualities: *Ad from a 1907* Canada Lumberman *praises the virtues of the Shay locomotive. [CL/RBL]*

to meet requirements not possible by any other type of engine." The Shay locomotive was modular, standard parts were mixed and matched. Each locomotive was designed to fit each company's needs, and no two were the same. Unlike the standard or "rod" locomotive, the Shay-geared locomotive had three pistons mounted vertically on its right side, just in front of the cab (They were sometimes known as "sidewinders"). These pistons turned a system of driveshafts (line shafts), slip-joints (sleeve couplings), universal joints (coupling rings), and gears which connected to and drove all the wheels on the locomotive. This unique design allowed the locomotive wheels to adjust to the bumpy and uneven track in logging operations, and allowed the Shay locomotive to pull heavy loads up steep grades.

The Shay locomotive came in two-truck and three-truck" versions. A truck consisted of four wheels on two axles mounted in a frame which was connected to the locomotive along the center line, allowing the frames to turn as the locomotive went around curves. The two-truck version consisted of a

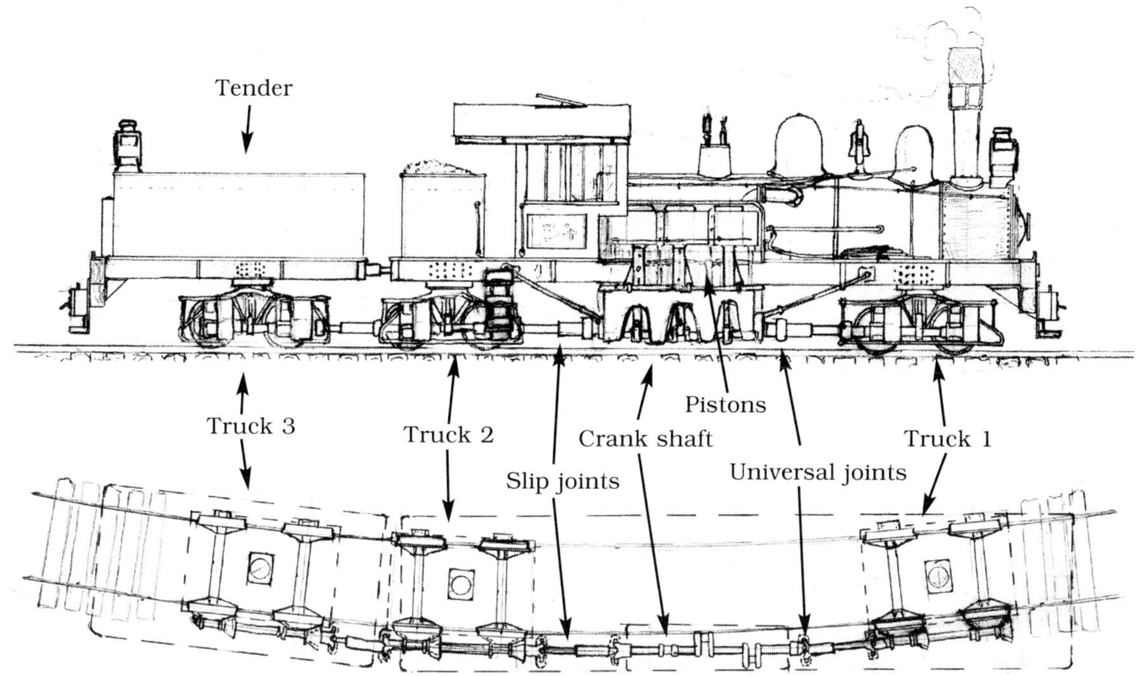

The "big engine" No. 54

[PM]

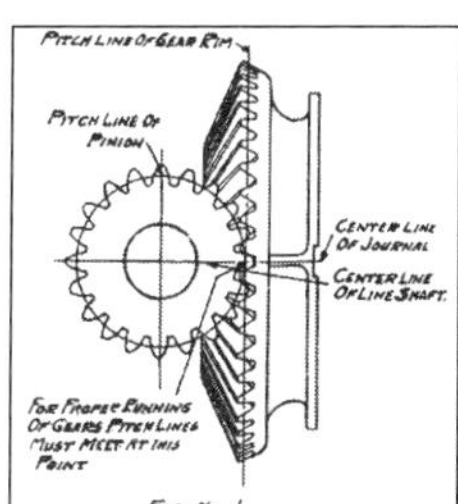

Turning: *A close up of the system of universal joints, slipjoints and gears that allowed the Shay to negotiate the curves on the railway. [UTA]*

Specifications for Lima Shay Geared Locomotive No. 1956 (Fassett No. 53)

The specifications below were provided by Lima Locomotive Works in response to a request by the Canadian Equipment Company when they purchased the used locomotive from the Standard Chemical Company in 1927. The letter was provided to us by the Allen County Museum in Lima Ohio.

Conditions of Operation:

Gauge of Track: 4'-8.5"
Weight of rail: Required 45
Maximum curve that the engine will take: 100' Radius.

General Dimensions:

Height, measured from top of rail: 14'-0"
Length, total wheel base: 39'-11.25"
Width: About 9'-6"
Cylinders, 3, diameter: 12", Stroke 15"
Driving wheels, number: Twelve, Diameter: 36"
Rigid wheel base: 4'-4"
Weight in average working order: 140,200 lbs.
Tractive power: 30,350 lbs.
Gear ratio: 1 to 2.25 (Pinions 20 teeth, rims 45)
Fuel capacity: 5 Tons coal.
Water tank capacity 3000 U.S. Gallons.
Boiler type: Extended wagon top, 46" Diameter.
Pressure 200 lbs. Fire box : 72 1/2 "X 42"

Other:

Jacket: Planished iron
Line shaft: front 4' dia., 2nd 5' dia., rear 4' dia.
Cab: Wood, rear entrance type.
Bell: 100 lbs.
Tires: 36' dia., 6' width, 3' thickness
Frames: Engine 12"- 40# I beam, Tender 10"

truck at the front and back of the locomotive. The three-truck version had a third truck, still connected to the drive shaft, supporting a tender behind the main locomotive.

Other distinguishing characteristics of the Shay locomotive were the offset boiler and the distinctive sound the locomotive made. Because the cylinders were mounted on the right side of the locomotive, the boiler was shifted to the left side of the locomotive, balancing the locomotive, but also giving it a very distinctive appearance. The Shay was a slow locomotive, climbing hills at 6 to 12 Km. (4 to 7 1/2 miles) per hour and on the level or downhill at only 15 to 30 km.(9 1/2 to 19 miles) per hour. Because it was geared, the cylinders fired rapidly and it sounded as if it were going 90 km. per hour.

A small announcement in the *Canada Lumberman* in November 1905, stated that "The Haskell lumber Co. of Fassett, Que. have recently purchased a new locomotive for hauling logs. It will be used on the Salmon River and Northern Railway." This was Shay locomotive No. 51. It was built in September 1905 and was a two-truck Shay rated at 60 tons weight and was around 15 meters (50 ft) long. This locomotive went to Fossmill in 1924 as their yard engine.

In 1905, the company also had a small Davenport saddle tank steam engine, No. 52. It was a 25-ton "rod" engine with only 4 drivers. Water was carried in a tank which wrapped around the boiler like a saddle. Stella McGibbon said that "There was no tender to it, a short bunty little thing, it just worked around the yard." In 1910, this engine was replaced by a similar locomotive from Montreal Locomotive Works , and was sold to Standard Chemical in 1924. The saddle tank locomotives and Shay locomotive were used to build the railway in Fassett, but after cutting began in 1907 a new locomotive was ordered. It was No. 53. and arrived in May 1908. It was a three truck Shay weighing 65 tons. This locomotive was eventually sold to Standard Chemical and then to the Canadian Equipment Company in 1926. It was scrapped in 1929. A third Shay was ordered in 1910 and arrived in October of that year. It was also a three-truck version, rated at 70 tons weight, and around 18 meters (60 ft) long. The three-truck locomotives, valued at $12,000 each, were the workhorses of the operation. "Strong enough to haul a good long load of logs" up the steep hills of the Laurentian and Algonquin Highlands. No. 54 was the main locomotive in the Fossmill operation.

One of the recurring problems with the locomotives and the log loaders was fires set by sparks from the engines. The Shay catalogue said that "The sparks [are] arrested by steel wire netting and a steel plate in the box until they are churned fine enough to pass through the netting." This method proved inadequate in Fossmill and led to a showdown with the Fire Service and Algonquin Park authorities. (see page 138)

Barnhart Log Loaders

Logging railroad operators used many methods to get the logs from the skidways up onto the log cars. Many were the same methods used for loading the sleighs in the bush, such as the "decking line" or "jammer". The Fassett Lumber Corporation used both of these methods, but mainly they used two steam-driven machines called "Barnhart Log Loaders". Valued at $4,500, these loaders were unique pieces of equipment especially adapted for logging.

Railway equipment historian Al Armitage drew a set of plans of the Barnhart loader to assist the Pennsylvania Lumberman's Museum in the restoration of the last surviving loader. He wrote in a series of articles in the *Narrow Gauge and Short Line Gazette* that "Some of the weird features of Barnharts may have resulted from the fact that they were not designed 'from scratch.' These curious machines were adaptations of a Marion steam shovel or derrick. In spite of their weird features, these monstrosities were a great success in the Eastern woods and many other parts of the U.S."

Henry Barnhart, co-founder of the Marion Power Shovel Company, Marion Ohio, developed the loader in 1887. Marion built 160 of the original model, and produced various other models until 1927. The Barnhart was the first loader in North America which could swing a full 360 degrees. It was able to pull itself from log car to log car as each was loaded or unloaded. An upright boiler in the rear of the cab powered a two cylinder steam engine that drove a continuously turning series of drums. The engineer used levers to engage and disengage the various chains and cables as needed. A large chain at floor level ran through a grooved ring and allowed the whole cab and 8 m. (25 ft.) oak boom to swing the full 360 degrees and then back again. This chain was uncovered and the engineer and fireman had to avoid it as the loader rocked and jerked back and forth as they worked. The Fassett film from 1934 shows the engineer looking through the front opening of the loader and working the levers, raising and lowering the tongs, and turning the loader as logs were lifted onto the log cars (see photo page 83 and video "Logging by Rail in Algonquin Park." See page 223).

The loader itself was mounted on a platform with 16 ten inch wheels which sat on two rails attached permanently to the top of the log cars. When the car in front of the loader was loaded, the crew attached a cable from the loader to the next car. By engaging this cable to the engine they were able to winch the loader over two U-shaped

The BARNHART STEAM LOG LOADER

A Logging System that is not an Experiment.

Tested by Time and its Efficiency proven by Economic Results.

Loaders built of weight and strength to suit any class of timber.

They work on rough or smooth track—on curves or grades.

Efficient — Full Particulars on Request — Simple in Operation

F. H. HOPKNIS & CO., MONTREAL

bridging irons onto the next car, so that the car it had previously been on could be loaded.

Log Cars

Fassett had 60 log cars valued at $400 each. They were made of heavy wood beams and fitted with rails on top so the loader could move over them. Logs were laid length wise on heavy cross beams resting over the pivots of the wheel trucks. When loading, the bottom outside logs were carefully placed to allow for as wide a load as possible (see page 83). This is why the Lima Locomotive Works recommended that logging railways be built in the wider standard gauge. Their catalogue said that "The bunks of logging cars for narrow gauge are shorter than for wide gauge, and logs must be piled higher than for wide gauge. For this reason standard gauge is usually preferable to narrow for logging." Logs were piled until the load was almost complete (each car could hold approximately 35,000 lbs of logs), and then chains were wrapped around to hold them in place. The final logs were then placed on top of the chains pulling them tight to the lower logs and locking them in place. Each log car was fitted with hand brakes.

Caboose

The caboose of the Fossmill railway is featured in many of the memories of those who worked at, lived in, or visited Fossmill. This small homemade caboose was mounted on a log car. It included a cupola with windows so the conductor and crew could watch the load. A small stove protected the train crew from the extreme cold of the Northern Ontario winters.

This little caboose carried many workers to the camps along the way, carried women and children out to the choice berry-picking grounds, and carried Park Rangers, Fire Rangers and the odd biologist into the Park. Fitted with hand brakes, it was often parked on a siding by means of a "running shunt", where it was uncoupled from the train while still in motion and, by flipping the track switch, sent off onto the siding and then stopped with the hand brake. It was during such a maneuver in 1931 that one of the most tragic incidents in the Fossmill story occurred (see page 142).

Other Equipment

The Fossmill railway also used flatcars for pulpwood, firewood and to carry supplies to the logging camps. Boxcars also carried supplies and were used to ship the finished lumber on the CNR. Hopper Cars, which had chambers which could be emptied from the bottom, were used to ship sawdust and chips on the CNR. A snow plow was also coupled to the locomotive to clear the tracks in the winter. In Fossmill the Model T Ford railcar used in Fassett was replaced with "speeders", small three-wheel gasoline powered machines which looked like a motorcycle with an outrigger to the opposite track.

The locomotives were fitted with steam brakes. For safety and efficiency, the cars attached to the locomotives on the main line railways also had automatic air brakes. But, on the Fossmill logging railway, because of the complicated shunting and maneuvering of log cars, boxcars, etc., automatic brakes were used only on the locomotive. The cars were connected to each other with "link and pin couplers," an oval shaped steel ring which slipped into a slotted bar at either end of the car with a steel pin put in place to stop it from coming out. This was more labour intensive than the automatic couplers on the main line railways.

The special qualities of Fossmill's Shay locomotives were demonstrated one day in the village when the crew of a large CNR freight locomotive miscalculated and got into trouble in the Fossmill yard. The CNR crew wanted to pick up gondola cars that were not placed in the correct position on the sidings close to the main line. Impatient, the engineer decided he would move them rather than wait for the company locomotives. John Gleason remembered, "It's quite a little slope there so he gets down there but then he couldn't get out, couldn't get going back up." Peggy Brushey remembers that "These vertical drive engines could take things up a hill where the CNR locomotive couldn't. The CNR engineer thought his engine could do anything — sounds like a kid's fairy tale. Down goes the engine. By this time everybody was alerted, the kids were out of school — to see this great big thing go down that spur and down to the millpond. It didn't go far before it was stuck. Then they thought 'if we just back up and take a run at it we're going to make it,' so they backed up and couldn't do it. After a couple of tries they took everything off. No matter what they did they could not get that thing up. So they had to get the company engines which just pushed that huge engine right up the grade onto the CNR. Wasn't that ironic? You could imagine what Margaret's father (the locomotive engineer) thought."

The locomotives and equipment of the main line railways were large, fast and able to pull long lines of railway cars over the miles of relatively level and even track. The locomotives and equipment used on the Fossmill railway were specially developed and adapted to the difficult conditions faced in the backwoods of Northern Ontario and Algonquin Park. The people of Fossmill were proud of their locomotives and loaders because, like themselves, they were especially adapted to life on the edge of civilization.

Appendix B: “We never saw the colour of a dollar bill” The voucher Relief payment system

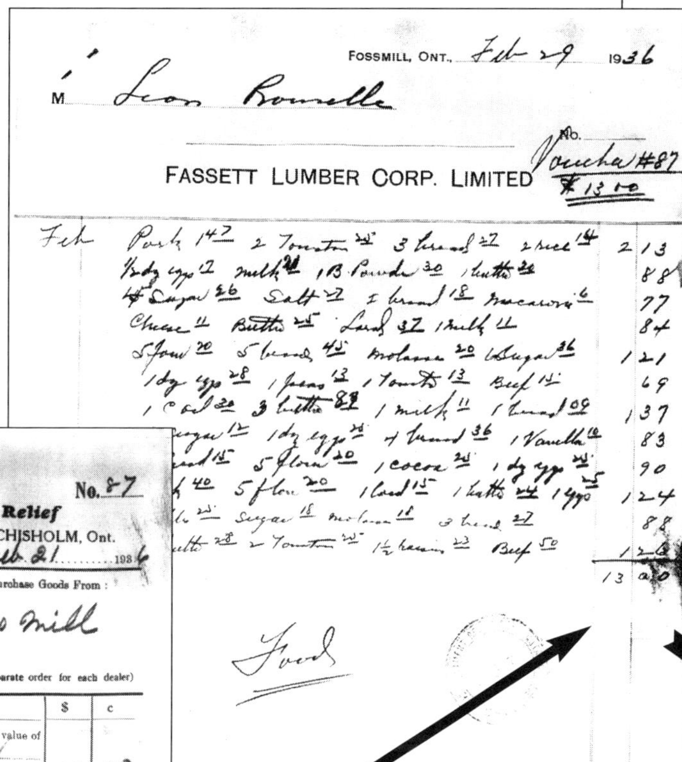

FOSSMILL, ONT., Feb 29 1936

M Leon Rouselle

FASSETT LUMBER CORP. LIMITED

No. Voucher #87 $13.00

Food

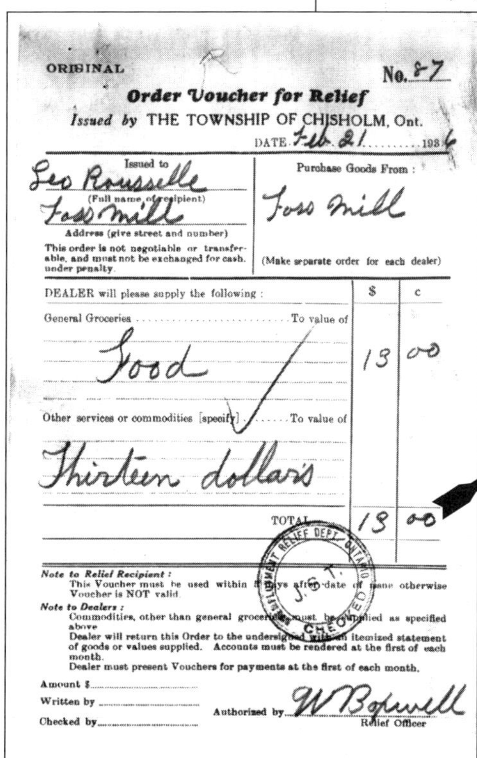

ORIGINAL

No. 87

Order Voucher for Relief

Issued by THE TOWNSHIP OF CHISHOLM, Ont.

DATE Feb 21 1936

Issued to: Leo Rousselle (Full name of recipient)

Foss Mill — Address (give street and number)

This order is not negotiable or transferable, and must not be exchanged for cash under penalty.

Purchase Goods From: Foss Mill

(Make separate order for each dealer)

DEALER will please supply the following:	$	c
General Groceries .... To value of — Food	13	00
Other services or commodities [specify] .... To value of — Thirteen dollars		
TOTAL	13	00

Note to Relief Recipient: This Voucher must be used within [illegible] days after date of same otherwise Voucher is NOT valid.

Note to Dealers: Commodities, other than general groceries, must be supplied as specified above. Dealer will return this Order to the undersigned with an itemized statement of goods or values supplied. Accounts must be rendered at the first of each month. Dealer must present Vouchers for payments at the first of each month.

Amount $ ....

Written by ....

Checked by ....

Authorized by W Boswell, Relief Officer

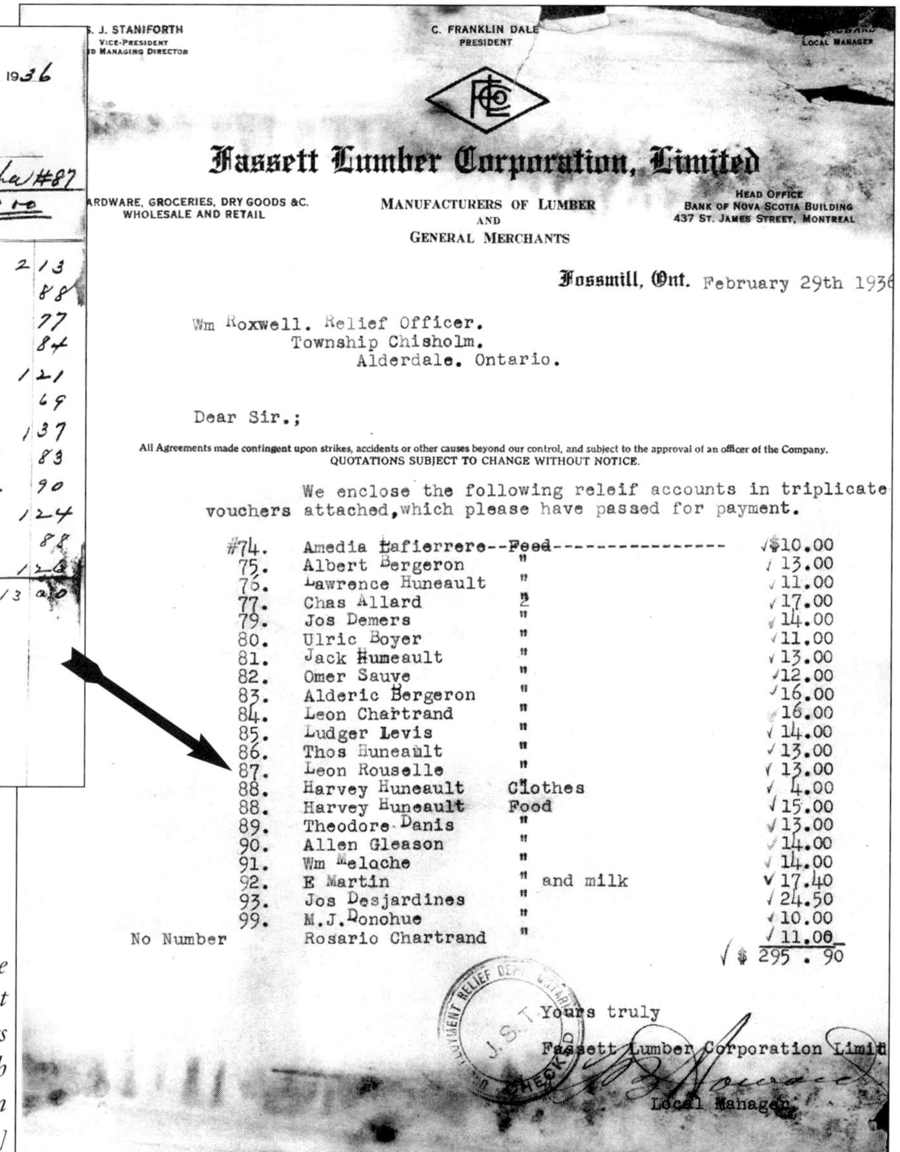

J. STANIFORTH, Vice-President and Managing Director — C. FRANKLIN DALE, President — Local Manager

Fassett Lumber Corporation, Limited

Hardware, Groceries, Dry Goods &c. Wholesale and Retail — Manufacturers of Lumber and General Merchants — Head Office: Bank of Nova Scotia Building, 437 St. James Street, Montreal

Fossmill, Ont. February 29th 1936

Wm Boxwell. Relief Officer.
Township Chisholm.
Alderdale. Ontario.

Dear Sir.;

All Agreements made contingent upon strikes, accidents or other causes beyond our control, and subject to the approval of an officer of the Company. QUOTATIONS SUBJECT TO CHANGE WITHOUT NOTICE.

We enclose the following releif accounts in triplicate vouchers attached, which please have passed for payment.

No.	Name		Amount
#74.	Amedia Lafierrere	Food	$10.00
75.	Albert Bergeron	"	13.00
76.	Lawrence Huneault	"	11.00
77.	Chas Allard	"	17.00
79.	Jos Demers	"	14.00
80.	Ulric Boyer	"	11.00
81.	Jack Huneault	"	13.00
82.	Omer Sauve	"	12.00
83.	Alderic Bergeron	"	16.00
84.	Leon Chartrand	"	16.00
85.	Ludger Levis	"	14.00
86.	Thos Huneault	"	13.00
87.	Leon Rouselle	"	13.00
88.	Harvey Huneault	Clothes	4.00
88.	Harvey Huneault	Food	15.00
89.	Theodore Danis	"	13.00
90.	Allen Gleason	"	14.00
91.	Wm Meloche	"	14.00
92.	E Martin	" and milk	17.40
93.	Jos Desjardines	"	24.50
99.	M.J. Donohue	"	10.00
No Number	Rosario Chartrand	"	11.06
			$ 295 . 90

Yours truly
Fassett Lumber Corporation Limited
Local Manager

***Relief:** Most of the men listed here went on to long careers with the Staniforth Lumber Company in Kiosk. [RMu]*

THE DENIS FAMILY
Notes on the cover photograph

The Denis family photo on the cover of the book captures the important reciprocal relationship between the people of Fossmill and the Fassett Corporation. The attractive young faces also raise the question of what went on in the years before and after the photo was taken.

The girl on the left is Rachel Denis. She married local boy Amedee Laferriere in 1931 at Fossmill. She was living in North Bay in December 1995 when she was interviewed and provided initial information on the family. She died in April 1996 and her son Gaetan, the family historian, provided additional information and direction. An interview with Claire (third from left) in June 1998 also provided additional information. The following is a brief profile of the struggle and survival of one of the Fossmill families.

Their father Albert N. (Polydor) Denis worked for the Fassett Lumber Company and lived in Montebello, Quebec with his wife, Eleanor. They had 6 children. Eleanor died during childbirth in 1923 at the age of 39. Albert was unable to care for the children: Rachel was placed in a convent, Claire and Octave in a orphanage and Cecile was "adopted" by another family, and Laurette and Juliette went to work as housekeepers for wealthy families in Rockcliff Park in Ottawa.

Albert came to Fossmill at its start to help build the mill. He also built a small house for himself and later helped build other houses and buildings, including the church. On July 1, 1927 (Canada's 60th Birthday) he brought Rachel, Claire and Octave from the Montebello area to live with him in Fossmill. The photo on the cover was taken about 2 years later when he brought the family together for one of their occasional visits. In the photo from left to right are Rachel (17), Cecile (7), Claire (13), Juliette (20), Octave (11) and Laurette (19). Rachel and Octave are now deceased. Cecile, Juliette and Claire now live in Ottawa and Laurette lives in Rapide Sept near Cadillac in Northern Quebec.

The following profile highlights Rachel, Octave and Claire who played a more significant role in the Fossmill story.

Rachel, as mentioned above, came to Fossmill at age 15 with her brother, Octave, and sister, Claire, to help her father and keep the family together. She did not attend school in Fossmill and had to learn English on her own. Things changed for her when she met local mill worker Amedee Laferriere. Amedee and Rachel were married in 1931. They had 2 children while at Fossmill and 4 at Kiosk. There was little accommodation at Kiosk but they were fortunate to be able to purchase the old Coristine railway section house nearby. Amedee's brother Octave later married and moved into the same house. Amedee drove a truck for the Staniforth Company for many years and had the important sawfiler's job in the mill. He is also remembered as an outstanding fiddler and as the town barber who gave thousands of hair cuts over 40 years. Their daughter, Vivian, was one of the first women to work at Kiosk when she started in the Staniforth Veneer Mill. Son, Gilles, and grandson, Guy, both work in the forest product business today at Columbia Forest Products in Rutherglen, continuing over 100 years and 4 generations of commitment to the trade.

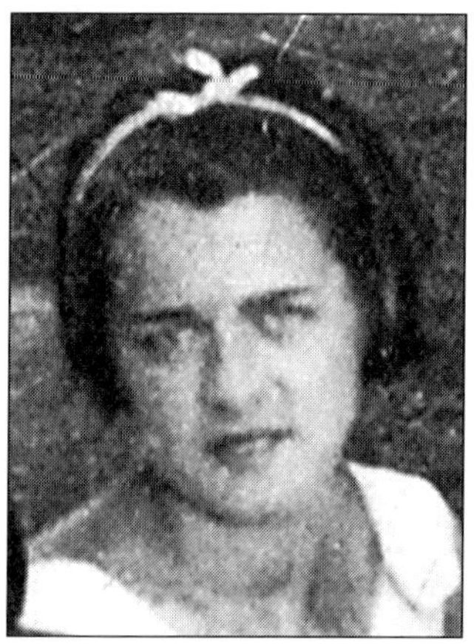

Claire was 13 years old when the photo was taken and was living in Fossmill with her father, brother, and sister Rachel. In a 1998 interview she recalled how much she liked Ted Priest, the teacher at the Fossmill school and how he taught her English for the first time, including strict rules of grammar. She talked about getting caught smoking on one occasion and getting the strap at the school and subsequently being punished at home as well. She recalled how her father lived for the job as many of the Fassett Company men did. Her father made bobsleds for the children and they would ice the slide and sled for hours. She was also friends with Irene Howard, the manager's daughter, and used to go to the movies in North Bay with Mrs. Howard and Irene. "Mrs. Howard for me was a real gem - so was Mrs. McGibbon". Her memories of her years in the orphanage were not as positive and she showed her disdain for the fact that she was poorly fed and had scurvy when she left. Feisty and articulate Claire expressed her opinion about the nuns in no uncertain terms. Claire contacted her sister in Ottawa when she left school at 15, and left Fossmill to join her as a housekeeper. She married Tony Grenier in 1930. Claire and her husband had 3 children and ran a unique dyeing business in Ottawa for years. She recalled discussions with some of her well known clients such as Mrs. Lester B. Pearson, Mrs. John Diefenbaker and Barbara Ann Scott.

Claire 1998

Octave went to school in Fossmill and worked in the mill. He later worked at Kiosk and joined the army from there. He served in the Aleutian Islands and in Europe and was wounded in action. He worked for the Department of Veterans Affairs in Ottawa and was married and had 3 children. Octave died in 1989.

Cecile was "adopted" by the Allard family in Sturgeon Falls but kept in touch with her biological family, including the visit when the photo was taken. She became a nurse and moved to Ottawa where she married John Porter and raised their 4 children plus 2 from John's previous marriage. She currently suffers from Alzheimer's and resides in a nursing facility in Ottawa.

Juliette, on her visits to Fossmill, wore the fashionable short skirts and hairstyles of the roaring 20s and drew admiring glances from the men of the village along with indignant stares from the churchgoing women. She did not marry and now lives in a seniors' facility in Ottawa.

Laurette was the "mother figure" to the other children and was admired and respected for the role she played. She worked as a housekeeper in Ottawa until she married Henri Gauthier and moved to Rapide Sept near Cadillac, Quebec where, at age 90, she enjoys her many pets.

Earl Orr with author Doug Mackey

Interviews

The following is a list of most of the formal recorded interviews or contacts conducted to write *The Fossmill Story* (An * beside the names of 9 interviewees indicates that the person is deceased as of the book going to press). There were many additional informal contacts that were helpful. The generous support of everyone we met is greatly appreciated and we regret that many excellent stories and photographs could not be used due to space and time restraints.

*1. **Anderson, Lloyd** – Powassan, 1995. Chisholm resident.

2. **Anderson, Myles** – Chisholm, 1996. Chisholm resident.

3. **Anderson, Richard "Dick"** – Powassan. 1995. Chisholm resident.

4. **Anderson, Stanley** – Powassan, 1995. Chisholm resident-former employee.

5. **Bergeron, Roger** – North Bay, 1995. Son of Fossmill Millwright Alderic Bergeron. Worker at Kiosk.

6. **Brushey, Margaret "Peggy"** – Powassan, Apr. 28, 1995, May 31, 1995, Jan. 12, 1996. Daughter of Jack and Stella McGibbon.

7. **Clark, Nora** – Hagersville, 1996. Daughter of Allan Gleason.

8. **Cooper, Ken** – South River, 1999.

9. **Cossar, Bernie** – Chisholm, 1996. Chisholm resident.

*10. **Coulas, Myrtle** – North Bay, 1996. Youngest child of Gordon and Myrtle MacDonald.

11. **Danis, Germaine** – North Bay, May 9, 1996. Daughter of Alderic Bergeron.

12. **Danis, Raymond** – North Bay, 1995. Child at Fossmill and Kiosk.

*13. **Demers, Louis** – North Bay, Nov. 24,1995, Dec. 13,1995, April 12, 1996. Worker at Fossmill and Kiosk.

14. **Demeza, Joe** – Belleville, 1995. Fossmill teacher.

15. **Donahue, Claudette** – Ottawa, 1998. First woman to work in Kiosk mill.

16. **Donahue, Gerry** –Grandson of Fossmill Railway Superintendent Morris Donahue.

17. **Dwyer, Bill** – Temagami (by phone), 1998. Son of Fossmill sawfiler Paddy Dwyer.

18. **Frederick, Margaret** – Powassan, Apr 18, 1995, Nov. 17, 1995. Daughter of locomotive engineer Allan Gleason.

19. **Gleason, Fran** – Written memories of Fossmill, 1996. Daughter of locomotive engineer Allan Gleason.

20. **Gleason, John** – Elmvale, 1996. Son of Allan Gleason. Worked at Kiosk.

21. **Grabowski, Margaret** – Powassan, 1995, Fossmill resident 1940's

22. **Grant, Ivy** – Barrie, 1996. Daughter of former

Louis died in June 1999, and fortunately, like several others opposite, left his Fossmill memories for us to share.

Louis Demers

Donald Staniforth

Margaret Frederick

Donat Huneault

John Rouselle

Roger Bergeron

Chisholm residents Charles Millar and Rose Hall.

23. **Grenier, Claire** – Ottawa, 1998. Daughter of Fossmill carpenter Albert Denis.
24. **Hendrickson, Barbara** – North Bay, 1999. Daughter of Fossmill CNR Station Agent Jack Scott.
25. **Huneault, Donat** – Mattawa, March 29,1996, Apr. 8, 1996. Mill worker at Fossmill and Kiosk.
26. **Koch, Otto** – North Bay, 1997. Manager at Kiosk.

*27. **Laferriere, Rachel** – North Bay, 1995. Wife of Fossmill and Kiosk employee Amedee Laferriere.

28. **Laferriere, Elise** – North Bay, 1996 Wife of Fossmill and Kiosk employee Oscar Laferriere.
29. **Laferriere, Gaetan** – 1996-99. Son of Rachel and Amedee.
30. **Laferriere, Guy and Gilles** – North Bay, 1996.Son and Grandson of Amedee.

*31. **Laferriere, Tom** – Powassan, Dec. 13, 1995, Feb. 26, 1996. Chisholm resident.

32. **Larochelle, Mike** – South River, 1995. Former manager at Kiosk.
33. **Legacé, Ila** – Mattawa, Oct. 24, 1995, Dec. 7, 1995. Daughter of Elva and Lawrence Huneault.
34. **Leyzac, Bernice** – Mississauga, 1996. Daughter of "Man" Martin, Fossmill Railway Brakeman.
35. **Litkee, Doreen** – Chisholm, 1996. Chisholm resident.
36. **Martin, Lionel** – Sudbury, 1996. Son of "Man" Martin, Fossmill railway brakeman.
37. **Maxwell, June** – Ottawa (by phone), 1996. Daughter of Roy Summers the paymaster for the Fassett Lumber Corporation.

*38. **McGibbon, Stella** – Powassan, 1995. Wife of Woods Superintendent Jack McGibbon, mother of Peggy Brushey.

*39. **Nault, Lois** – Sudbury, 1996. Daughter of Joe Donahue.

*40. **Nault, George** – Sudbury, 1996. Kiosk worker and son-in-law of Joe Donahue.

41. **Orr, Earl** – Toronto, 1996. Son of stationary Engineer John Orr.
42. **Donahue, Ruth** – Pembroke 1996 . Daughter of Fossmill stationary engineer John Orr.
43. **Pilgrim, Muriel** – 1996. Wife of early Fossmill railway fireman.

*44. **Priest, Ted** – Stratford (by phone), 1995. Fossmill teacher.

45. **Rouselle, John** – North Bay, 1995. Father worked in cookery-former employee, Kiosk.
46. **Sliwka, Edward** – Ottawa (by phone), May 1999. Son of Fossmill machinist.
47. **Smith, Max** – Chisholm, 1995/99. Chisholm resident.
48. **Staniforth, Donald** – Toronto, 1996. Son of Sydney Staniforth.
49. **Staniforth, Robert** – Florida (by phone), 1998. Son of Sydney Staniforth.
50. **Summers, Roy** – North Bay 1998. Grandson of Roy Summers the paymaster for the Fassett Lumber Corporation.
51. **Tennant, Lloyd** – Chisholm, May 6, 1995. Son of logger George Tennant.

John Gleason with author Paul Mackey

Ila Legacé

Nora Clark

Doreen Litkee

Max Smith

Peggy Brushey

Stanley Anderson

Glossary

The following are definitions of some of the terms used in the text as indicated by **bold** type. Terminology sometimes varied from location to location, so some terms may have different meanings elsewhere than those indicated here. Many words also have different meanings in other contexts. For example, the men in the bush are called various names – bushworkers, woodsworkers, lumberjacks, loggers and, more recently, forest workers.

Annual allowable cut – The amount of timber that is permitted to be cut to keep the forest at the same level of growth in a *sustained yield* logging system.

Barnhart steam log loader – A steam shovel adapted for loading logs (see page 206).

Beaver – Bushworker who cleared skidding and hauling roads.

Binding Chains – A chain that was wrapped around a load of logs. Usually the last few logs were placed on top of this chain to pull it tight.

Board foot – A unit of measurement of dimensional lumber 1inch thick x 12 inches wide x 12 inches long. Wholesale lumber is usually sold in thousands of board feet.

Boom – A line of logs chained together at each end to encircle and control floating logs.

Brakee – The colloquial name used for the man who operates the brakes on log hauling sleighs and log cars to control speed.

Buck – To saw a felled tree into log lengths.

Buck beaver – The person in charge of a road crew building a hauling road. The workers are called beavers.

Bunk – Timber crossbars forming the bed for logs in a logging sleigh. Also the bed people slept on at night.

Bush Walker – colloquial term for the *woods superintendent* of a logging company. Also see *Cruising*.

Cadge – To deliver supplies usually with a cadge team on a cadge road in the bush.

Calk Boots – High cut boots with ¼ inch spikes in the sole, used when walking on logs in slippery or risky situations.

Caterpillars – A brand name type of tractor with a linked tread to provide better traction. Also a generic name for tractors using linked treads.

Chickadee – The person, often a youngster, responsible for removing manure and keeping a hauling road clean and smooth for the sleighs passing over them.

Clearcutting – The process of cutting all trees from an area without regard for the regeneration or sustainability of the area.

Cull – To remove poor and defective logs or lumber from a pile. These logs were called "cull logs" and the lumber "cull lumber".

Cupola – An observation tower projecting above the roof of a caboose so that the conductor and brakemen can watch the train.

Cord – Pile of wood 4 feet by 4 feet by 8 feet.

Cordwood – Small logs cut to 4 foot lengths.

Cookee – A cook's helper who does whatever is necessary to prepare the camp meals.

Cross-chain – The chain going from one inside corner of a sleigh bunk across to the other bunk to allow the bunks to operate separately on rough roads.

Cruise – To walk over an area to be logged and estimate the amount and value of the standing timber.

Decker – See *decking*

Decking – The temporary stacking of logs in a pile or skidway. The men who deck are called *deckers*, and a *decking line* and horse are used to assist in the work.

Decking line or chain – A chain or rope with a hook on one end that passes around a log to a pulley and then to a horse so that when it is pulled by the horse the log rolls onto a skidway or a log sleigh.

Depot – The main headquarters for a logging operation.

Diameter limit – An established appropriate minimum size below which logs will not be cut.

Dogger – Mill worker who stands on the saw carriage in the mill and positions the *dogs*.

Dogs – Pointed metal tool used to hold a log in place for cutting.

Edger – A machine with a number of stationary and adjustable saw blades that cuts the rough edges from boards to straighten the edges.

Endless-flat-link Chain – A conveyor belt made of spiked steel links that ran up the centre of a *jackladder*. The spikes grabbed the logs at the bottom and pulled them into the mill.

Flywheel – The large drive wheel on a steam engine in a mill power plant from which power is taken by a system of belts, pulleys and shafts.

Gas car – A small open flat four wheel railway car used for transporting people and goods, or for pulling cars without power. Some are hand pumped and others are driven by gasoline engines. (also see *speeder*)

Go-back road – A road used by a team and sleigh returning to a skidway so as to not interfere with loaded sleighs on the main hauling road.

Gondola car – A large railcar with a full length storage tank for sawdust or chips with an opening in the top for filling.

Grade – The name for the quality of a piece of lumber as established by a *grader* so all lumber purchased is of similar quality.

Grade (2) – The extent that the rails on a railway or the ground itself is off level.

Grader – A worker who determined the *grade* of each piece of lumber.

Hauling road – A road built specifically for the safe and easy hauling of logs.

Hog – A mill machine that grinds or chews scrap wood into chips for burning in the firebox of a boiler.

Hot pond – A log pond that was heated in the winter.

Incendiary fire – A fire that is purposely set; better known as an arson fire.

Jackladder – An angled ramp that has an *endless-flat-link chain* moving up the middle to hook and carry the logs to the top of the ramp and into the mill.

Jammer – A tall A-frame of two logs mounted on a sled and rigged with a pulley and rope to lift logs as required. It is sometimes worked by horses and sometimes with steam power.

Jobbers – Contractors who takes on the job of logging an area for a lumberman.

Jumper – A lumberjack who moves from camp to camp either because they find the work too hard or the conditions unsatisfactory.

Kicker – The device that pushes or kicks a log from the trough at the top of a jackladder onto the logdeck.

Lath – The 3/8 inch x 1 ½ inch x 4 foot strips of wood used on walls prior to plastering before drywall was developed.

Lath mill – Machinery that cut slabs and edgings into *lath.*

Live rollers – Continuously moving rollers that automatically pulled slabs away from the bandsaw.

Log deck – The platform on which logs are placed prior to being cut into lumber by the bandsaw

Log dump – Central log pile

Logging camp – A base site for a logging operation providing accommodation, meals and related activities, as required by the operation.

Logging limit – A section (berth) of government owned land leased to a lumber company for the removal of timber. Also sometimes used as a generic term for the area a company is logging.

Logger – A generic term for a person cutting logs. In eastern Canada usually called a lumberjack.

Lorry – A platform with 4 wheels used to transport goods on a railway.

Lumberman – Up until 1850 anyone engaged in the lumber trade, with little distinction between the owner and the worker. After 1850 the owner became known as the lumberman and the workers as lumberjacks.

Mantle – A fine bag fixed around a gas flame in a lamp to produce a bright glow.

Merchantable timber – A log that can be profitably made into lumber.

Peavey – A long-handled tool with a pointed end and hinged hook, used to roll or turn logs.

Pickets – Pieces of lumber cut from slabs of wood cut from logs and used in the manufacture of furniture requiring small pieces of lumber.

Pulpwood – Any wood with a fibre suitable for making pulp for paper production.

Railcar – An automobile whose wheels have been adapted so that it can travel on railway tracks.

Reciprocating shotgun carriage – The steam driven platform on wheels that carries a log into the bandsaw blade for cutting. It shoots forward and back on a track, cutting stock on each forward motion. It carries a *dogger* and *setter* who lock and adjust the log to facilitate the cutting.

Resaw bandsaw – The saw used to cut planks into thinner boards of specific sizes to fill orders at the mill.

Roller – The bushworker responsible for rolling logs up skids onto a log pile in a skidway or depot using a *peavey* and assisted by a *decking line.*

Roller (2) – A portable platform used in the lumberyard to transfer lumber from the tramway over a first pile of lumber to a second pile.

Running shunt – The technique of uncoupling a car from a train in motion so that the car can roll freely onto a siding.

Saddle tank switching engine – A small locomotive used to move railway cars in a yard. Water was carried in a tank that wrapped round the boiler like a saddle.

Sandpiper – The bushworker who prepares and applies sand to a hill on a hauling road to slow sleighs down.

Sawfer's loft – A room above the main sawing floor where the saws were sharpened and maintained.

Saw log – A log suitable for sawing into lumber as distinguished from *squared timber* or *pulpwood.*

Sawyer – Worker who cuts down trees in the bush and the worker who saws logs in the mill.

Scaler – Person who measures logs to calculate their volume and value. He works with a partner who measures the opposite end of the log.

Scaling stick – A rule, marked with a table, used by the scaler to quickly determine the volume of wood in a log.

Selection cutting – The removal of mature, defective, or dead trees to allow young trees to mature naturally for future use.

Sender – The bushworker who guides logs when *decking* and also the person who hooks the logs to be lifted and lowered onto a log car.

Setter – The mill worker on the shotgun carriage responsible for adjusting the width of the cut in a sawmill at the direction of the head sawyer.

Shay-geared locomotive – A specially geared engine that allows the pulling of heavy loads on steep grades.

Shunt – To move railway cars onto sidings, see *running shunt.*

Skids – Sturdy poles placed at right angles to a log pile in order to roll logs onto the pile.

Skidding – The process of hauling logs from where they are cut, along skidroads or trails, to a skidway.

Skidway – temporary log pile close to the cutting

area where logs are stored until hauling roads are frozen.

Slab – The rough outside piece cut from a log in the process of sawing the log.

Slash – The branches cut from trees or the small trees removed when clearing trails.

Snaking trail – A log-removal trail that zig-zags down a *grade* to prevent accidents caused by the acceleration of the team of horses.

Sorter – The mill worker who separates different grades and sizes of lumber for storage.

Spark arrestor – A mesh enclosure on the stack of an engine to prevent sparks from escaping and starting a fire.

Speeder or velocipede –A small three wheel railway car, which was easily removed from the tracks.

Square timber – A large piece of timber squared by a crew using broad axes and other tools in the early days of logging.

Sprinkler tank – A sleigh with a wooden box mounted on it to hold water that is spread on hauling roads to make them stable and smooth to accommodate hauling sleighs.

Steam log-turner – The device used to rotate a log so another cut can be made when sawing into lumber. See note 11 bellow.

Stickers – The long narrow pieces of wood used between layers of piled lumber to allow for air circulation in the drying of lumber.

Stumpage – The royalty paid by a lumberman to the government for the logs cut on a site.

Sustained yield – The concept of cutting logs in such quantity as to allow the forest to continue to grow naturally or be replanted, pending a similar cut in the future.

Swampers – Road builders, see *beavers* and *buck beavers.*

Switchback – A rail line that zig-zags down the side of a hill in order that the track is not too steep for the locomotive to pull the loaded cars.

Tailer – Bushworker who controls the horse that pulls the *decking line* when loading log sleighs.

Tally Board – Clipboard used by the *scaler* to record the volume of logs.

Tanbark – The bark of the hemlock tree used in the tanning and preserving of leather.

Teamster –The person in charge of a team of horses.

Timber – Trees that have a merchantable value to a lumberman.

Top loader – The bushworker who controls the building of a pile of logs by standing on the top of the pile and directing logs into place.

Tote road – A temporary rough road used to get goods to and from the bush. Sometimes referred to as a *cadge road.*

Two-man crosscut saw – Long saw with handles at both ends used in felling and *bucking* trees.

Trail cutter – The bushworker who clears brush to create a trail for the removal of logs by a teamster.

Tram – Narrow gauge rail car used to carry lumber.

Tramway – The elevated framework and track in a lumberyard used for transporting lumber into the lumberyard.

Transfer chains – General name for a number of different types of continuously moving chains that moved lumber and slabs to various parts of the mill.

Transfer track – The lateral moving bed that moves tram cars into alignment with the tracks on a tramway so tram cars can progress in to the yard.

Undercut – A notch cut in the tree before sawing it ensured the tree fell in the direction wanted.

Van – The store in a camp where men purchased items they needed for daily use.

Velocipede – See speeder

Walked – See cruise

Woods Superintendent – The man in charge of a company's forest operation. Sometimes called a bush boss, or walking boss.

Notes

1. (throughout) The *Canada Lumberman* was a trade newspaper published for decades beginning in 1880 in various forms, monthly, weekly, and under various names. It is a rich source of news, lore, and education about the logging, lumbering, and retail lumber industry in Canada.
2. (throughout) The diamond shaped Fassett Lumber Company logo used as a divider was based on the iron stamps used to mark the end of each lumber companies' logs in the river drive era.
3. (page 33) See Brian D. Westhouse's article "*Trout Creek Logging Railway"* and column by Hartley Trussler in the *North Bay Nugget* June 14, 1974 (see Bibliography).
4. (throughout) *Our Chisholm Story* is a two volume history of Chisholm Township compiled and produced by the Chisholm Women's Institute.
5. (page 60) The report on the Fassett logging operation around Lake Manitou was written by three University of Toronto forestry students: A. S. L. Barnes, A. W. Leman, and G. R. Sonley, in December 1928. They visited three of the six Fassett camps and based most of their information on the work of contractor, Leon Lafleur from Lachute, Quebec.
6. (page 62) Ian Radforth's book *Bushworkers and Bosses* is an excellent study of the history of the conditions of the bushworkers and the effects of technology in the Ontario logging industry.
7. (page 66) The number of trees cut by each team of lumberjacks or by each camp varied widely depending on the number of lumberjacks in a team, the size of trees, weather conditions, etc. The figures here were determined from figures of average board feet per acre, average board feet per

log, and average logs per tree in the Fossmill forestry student report.

8. (page 70) The study was done by Alexander Koroleff in 1931. See page 97 of *Bushworkers and Bosses (see bibliography).*
9. (page71) For analysis of "jumpers" see page 7 and 39 of *Bushworkers and Bosses.*
10. (page 86) The interview with Dr. MacLulich was done as part of an oral history project of Algonquin Park in 1976. He was interviewed by Ronald Pittaway.
11. (page 91) In the 1920s and 30s the steam log turner was referred to in quotes as the "nigger" or simply as Nigger in the industry. Workers interviewed were reluctant and embarrassed to use the term, some using "kicker" (another piece of equipment) instead. We have used the technical and literal, name "steam log turner".
12. (page 98) The prevalence of static electricity in the bowels of mills was discussed on page 72 of the book *Company Town (see bibliography).*
13. (page 102) The Denis family biography was written by Gaetan Laferriere the son of Rachel and Amedee Laferriere.
14. (page 137) This *North Bay Nugget* report on the Moose Lake fire appeared in the spring of 1930. It was written as though the fire had occurred in the spring, but the details in the Fire reports for the fall of 1929 match the news report, while the fires in the spring of 1930 do not. It appears that the reporter used information from the Fire Service for 1929 and reported it as having happened in 1930.
15. (page 139) The Cleveland-Sarnia Sawmill Company forestry student report was written in 1911 by W. L. Scandrett. It is available at the University of Toronto Archives.

Selected Bibliography

Books

Berton, Pierre. *The Great Depression 1929-1939.* Toronto, Ontario: McClelland and Stewart, 1990.

__________. *The Dionne Years: A thirties melodrama.* Toronto, Ontario: McClelland and Stewart, 1977.

Bondy, R. J. and W. Mattys. *Canadiana Scrapbook: The Confident Years -- Canada in the 1920's.* Scarborough, Ontario: Prentice-Hall of Canada, Ltd., 1978.

Braithwaite, M. *Canada's Illustrated Heritage: The Hungry Thirties (1930-1940).* Toronto, Ontario: Natural Science of Canada Ltd., 1977.

__(no author given)__. *Canada's Illustrated Heritage: The Crazy Twenties (1920-1930).* Toronto, Ontario: Natural Science of Canada Ltd., 1978.

Broadfoot, B. *Ten Lost Years, 1929-1939: Memories of Canadians Who Survived the Depression.* Toronto, Ontario: Doubleday Canada Ltd., 1973.

Brown, Ron. *Ghost Towns of Ontario. Vol. 2.* Toronto, Ontario: Cannonbooks, 1984.

Brozowski, R.S., D.L. Rees and K.H. Topps. *Bibliographic Guide to North Bay and Area.* North Bay, Ontario: Nipissing University College, 1979.

Choquette, R. *Language and Religion.* Toronto, Ontario: University of Toronto Press, 1975.

Feirer, J. L. *Cabinetmaking and Millwork.* Peoria, Illinois: MacMillan/McGraw-Hill, 1988.

Friends of Algonquin Park. *Logging History in Algonquin Park.* Algonquin Logging Museum, 1996.

Gould, R. *Calvin Remembers One Hundred Years, 1887-1987.* Calvin Twp., Ontario: Calvin Township Centennial Committee, 1987.

Gunning, C. *North Bay, the lean years: a decade of adversity.* North Bay, Ontario: Bond Printing and Graphics, 1996.

Kirton, Everett. *History of North Parry Sound District.* Nipissing, Ontario: Nipissing Museum, 1962.

__________. *Logging in Northern Parry Sound District.* Nipissing, Ontario: Nipissing Museum, 1977.

Koch, M. *The Shay Locomotive: Titan of the timber.* Denver, Colorado: World Press, 1971.

Leatherdale, M. Nipissing From Brule to Booth. The North Bay and District Chamber of Commerce. New Liskard, Ontario: Temiskaming Printing Co. Ltd., 1975.

Lima Locomotive Works Inc.. *Shay Instruction Sheets.* Parsons, West Virginia: McClain Printing Company, 1975 (Reprint).

Lima Locomotive and Machine Company. *Instructions for the care of Shay Geared Locomotives-1906.* Parsons, West Virginia: McClain Printing Company, 1972 (Reprint).

__________________________. *Repair Parts for Shay Geared Locomotives,* Parsons, West Virginia: McClain Printing Company, 1969 (Reprint).

______________________________. *Shay patent and direct locomotives, logging cars, car wheels, axles, railroad and machinery castings (catalogue).* Parsons, West Virginia: McClain Printing Company, 1971 (Reprint).

Lower, A.M. *The North American Assault in the Canadian Forest.* Toronto, Ontario: Ryerson, 1938.

Macfie, John. *Parry Sound Logging Days.* Erin, Ontario: Boston Mills Press, 1987.

MacKay, Donald. *Heritage Lost: The crisis in Canada's forests,* Toronto, Ontario: Macmillan of Canada, 1985.

_____________. *The Lumberjacks.* Toronto: McGraw-Hill Ryerson Ltd., 1978.

MacKay, N. *By Steam Boat and Steam Train: The story of the Huntsville and Lake of Bays Railway and Navigation Companies.* Erin, Ontario: Boston Mills Press, 1982.

_________. *Over the Hills to Georgian Bay: The Ottawa, Arnprior and Parry Sound Railway.* Erin, Ontario: Boston Mills Press, 1981.

MacKay, R. and Reynolds, W. *Algonquin.* Toronto, Ontario: Stoddart Publishing Co. Ltd., 1993.

Mennill, Paul. *Canadiana Scrapbook: The Depression Years – Canada in the 1930's.* Scarborough, Ontario: Prentice-Hall of Canada, Ltd., 1978.

Morel, L. *Mattawa: The Meeting of the Waters.* Mattawa, Ontario: Mattawa Historical Society, 1980.

Natural Resources Canada. *The State of Canada's Forests, Learning from History.* Ottawa, 1996-1997.

Neering, Rosemary. *The Depression.* Markham, Ontario: Fitzhenry and Whiteside, 1984.

Peterson, K. C. *Company Town.* Pullman, Washington: Washington State University Press, 1987.

Radforth, Ian. W. *Bushworkers and Bosses: Logging in Northern Ontario, 1900-1980.* Toronto, Ontario: University of Toronto Press, 1987.

Saunders, A. *Algonquin Story.* Toronto: Ontario Department of Lands and Forests, 1946.

Steer, W. *North Words: The Near North's History.* North Bay:1990.

Swift, J. *Cut and Run: The Assault on Canada's Forests.* Toronto, Ontario: Between the Lines, 1983.

Taim, A. *Almaguin: A Highland History.* Toronto, Ontario: Natural Heritage/Natural History Inc., 1998.

The Chisholm Women's Institute. *Our Chisholm Story, Vol. 1 (1880-1960) and Vol. 2 (1880-1974).*

Thompson, J. H. Canada 1922-1939: Decades of Discord. Toronto, Ontario: McClelland and Stewart, 1986.

Fassett, Quebec. *1918-1993 75 Fassett.* Fassett, Quebec: 1993

Articles and Periodicals

Coons, C. F. *"The John R. Booth Story"* In Your Forests Vol. 11, No. 2, Summer 1978.

The Beacon Herald, (Stratford, Ontario) January 12 1998.

Canada Lumberman, (Toronto) 1905-1936.

North Bay Nugget 1924 to 1936 and 1942.

The *Journal* article, Courtesy of Patricia Logan.

Churcher, C. *"The Salmon River & Northern Railway"* Ottawa, Ontario: Branchline, May 1988.

Westhouse, B. *"Trout Creek Logging Railway"* Ottawa, Ontario: Branchline, October 1990.

Armitage, Al *"Barnhart Log Loaders (Part 4)"* Los Altos, California: Narrow Gauge and Short Line Gazette, May/June 1994.

Gatke S. R. *"The Cascade Lumber Company (Part 2)"* Hillsboro, Oregon: TimberTimes, August 1998.

Internet

Fournier, R. D. (1996) Text of and Information on Regulation 17 (Ontario -1912) from Inter.Canada web page: www.naccess.com/~inter.canada/doc/reg-17e.htm

Information on Barnhart Log Loaders from Historical Construction Equipment Association, www.bigtoy.com/photo/rail_marion3.html, and the Pennsylvania Lumber Museum, www.lumbermuseum.org

Government Documents

Algonquin Park Technical Bulletin No. 10. *Names of Algonquin: Stories Behind the Lake and Place Names In Algonquin Provincial Park.* Whitney, Ontario: The Friends of Algonquin Park, 1993 (revised).

Algonquin Park Technical Bulletin No. 8. *A Chronology of Algonquin Provincial Park.* Whitney, Ontario: The Friends of Algonquin Park, 1993 (revised).

Algonquin Provincial Park Master Plan. Ontario Ministry of Natural Resources, 1974.

Chapman, L.J. and D.F. Putnam. *The Physiography of Southern Ontario*, Third Edition. Ontario Geological Survey, Special Volume 2. Ontario Ministry of Natural Resources, 1984.

Unpublished and Archival Sources

Algonquin Park Superintendent Reports, 1929-1934. Algonquin Park Museum Archives.

Chayer logging record books. Courtesy of Marlene Chayer.

Chisholm Board of Health minutes. Chisholm Township Archives.

Chisholm notes. Chisholm Township Archives.

Chisholm scrapbook. Chisholm Township Archives.

J. Sloat Fassett information. Courtesy of Bradford County Historical Society, U.S.

Faculty of Forestry logging reports, University of Toronto Archives

Bentley A. W. and Clarke T. A. *Report on the Logging Operations of the Fassett Lumber Company, Limited at Fassett, Quebec, December 1919.*

Barnes A. S. L., Leman A. W. and Sonley, G. R. *Report on Logging Operation of the Fassett Lumber Company, Limited conducted at Fossmill, Ontario, december 1928.*

Heimburger, C. and Kelly, T. *Report on Logging Operation of the Standard Chemical Company at South River 1926-1927.*

Jenkins F .T. and Brodie J. A. *J. R. Booth Lumber Company, Kiosk 1922-23.*

Nieuwejaar, O. *Report on Logging Operation of the Fassett Lumber Company, Limited conducted at Fassett, Quebec, December 1916.*

Scandrett, W. L. *Report on Logging Operations of Cleveland-Sarnia Sawmills Co. Conducted at Diver, Ont. 1910-1911.*

Forest Fire Reports 1923 to 1934, Archives of Ontario.

Lima Locomotive Works specification sheet. Allen County Historical Society.

Maclulich, D. Interview by R. Pittaway, November. 23, 1976. Algonquin Park Museum Archives.

Movie film of Fassett Lumber Corp. logging and mill operations, 1934. Courtesy of Donald Staniforth.

School daily registers 1930-34, Chisholm Township Archives.

Staniforth, S. J. personal record books. Courtesy of Robert Staniforth.

Timber License records, Archives of Ontario.

Photograph and Illustration Credits

The initials in front of each contributor follow the caption of each photograph they have contributed.

Individuals:

[AP] Ann Pike
[BL] Bernice Leyzac
[CG] Claire Grenier
[DA] Richard "Dick" Anderson
[DF] Doug Frederick
[DL] Doreen Litkee
[DM] Doug Mackey
[DS]Donald Staniforth
[GL] Gaetan Laferriere
[GM] Gilbert Michauville
[IL] Ila Legacé
[JD] Joe Demeza
[JE] Jim Edington
[JM] John Macfie
[KC] Ken Cooper
[MC] Myrtle Coulas
[LD] Louis Demers
[LG] Lucille Gagnon
[LN] Lois Nault
[MF] Margaret Frederick
[MG] Margaret Grabowski
[ML] Mike Larochelle
[NC] Nora Clark
[PB] Margaret "Peggy" Brushey
[PL] Patricia Logan
[PM] Paul Mackey
[RB]Roger Bergeron
[RD] Raymond Danis,
[RF] The Ralph Friske Collection
[RMu] Ray Muttart
[RM] Rose Maloche
[RS] Robert Staniforth
[JT] Jack Trussler
[TP] Ted Priest
[WL] Wayne LeBelle

Institutions:

[AO] Archives of Ontario
[CT] Calvin Township
[CWI] Chisholm Woman's Institute
[GA] Glenbow Archives, Calgary, Canada
[LMu] Latchford Logging Museum
[NAPL] National Air Photo Library
[ON] Ontario Northland Transportation Commission Archives
[RBL] Thomas Fisher Rare Book Library
[UTA] University of Toronto Archives
[UTL] University of Toronto Library

Publications:

[CL] Canada Lumberman
[NBN] North Bay Nugget

Index

For acknowledgements, bibliography, glossary, interviews, maps photo credits, sidebars, and places to visit see table of contents. Pages in **bold** indicate photograph or illustration.

A:

Accidents, 26, **27**, 69, 73-74, 97-98, 129, 142-143, 184-185
Alderdale, 11, **37**, **39**, 181
Algonquin Park, 9, 30-31, **32**, 42, 60, 139, 148, 171, 185, 195, 200-201
Alligator logging tugboat, **35**, **173**
Automobiles, **48**, 111-112, **111**, **112**, **113**, 182

B:

Barges, 18, **27**
Barnhart log loaders, **21**, **23**, **30**, 82-83, **83**, **137**, 157, **198**, **202**, 206-208, **207**
Boarding house, **45**, **48**, 49, **49**, **57**, **108**, 113, 145
Booth J. R., 9, 35, 42, 59, 60,171, 172
Booth Lumber Company, 42, **53**, 175, 201

C:

Caboose, 82, **84**, 86, **158**, 208
Canada Lumberman, 16, 17, 18, 25, 30, 41, 56, 71, 82, 85, 99, **115**, 143, 145, 149, 172, 206
Canadian National Railway, 37-39, 41, 54, 60, 95, 138, 152-155, 161, 167, 170, 171, 172-173, 177
  the mixed, 167, 177, 183, **184**, 193
Canadian Pacific Railway, 15, 17, 172
Chayer, Joe, 36, 63
Chisholm Township, 31, 35-39, **53**, 132-134, 148-149, 166, 193
Chiswick, 11, 56, 192
Christmas, 54, 68, 69, 71, 89, 113, 129, 146-147, 170, 177, 184, 188
Church,
  Catholic, 56, **56**, **101**, 131, 146, 168, 188-189, **194**, 195
  Protestant, 56, **130**
Company store, **45, 49**, 49, **57**, 113, **114**, 133
Crime, 129, 146-147
Cruising, 30, 59-60

D:

Dale, Charles, 43, 110, 161
Dances and celebrations, 125-126, 170
Depression, 11, 135, 141-142, 143, 148-149, 152, 160, 165-166, 170
Dating, 112, 125-126, 132, 133
Dionne quintuplets, 11, 152
Doctors, 111, 114-115, 129, 146, 147, 175, 184-85, 187, 188

F:

Fassett Lumber Corporation/Company, 13, 17-18, 31, 42-43, 60, 69, 72, 134-135, 139, 149, 163, 179, 192
Fassett, J. Sloat, 16, **16**, 17, 42
Fassett, Quebec, 13-29, **29**
Finnish workers, 63, 65, 70-71, 80, 151-152
Fire rangers, 122, 137-140, **139**, 157, 167
Fires,
  forest, 137-141, **138**, **166**, 167
  mill and yard, 12, **12**, 144-145, 155-159, 199-200, **201**
Food and shopping, 106, 113-114, 142, 167, 170, 182-183
Forestry student reports, 16, 60, 61, 63, 70-71, 138, 160
Foster, William, 31, **31**, 33, 36, 41, **41**, 60
Foster's Mill, 42, **42**, 47, **53**, **57**
Free Grant and Homestead Act, 9
French/English relations, 52, 102, 115, 129-132, 168
  regulation 17, 52, 131

G:

Gravelle, Father Joseph, 56, 131, 143, 146, 147, 168, 179, 189

H:

Hardwood, 19-20, 30, 31, **41**, 42, 59, 92, 171, 175, 182
  Canadian Hardwood Bureau, 149, **149**
Haskell Lumber Company, 15
Health, 117, 129, 146, 147
Hemlock trees, 75, 96
Hobos, 152-154, 167-168, **168**, 181-182
Horses, **25**, **62**, 63-64, 75, **75**, 146
Hot pond, **23**, **46**, 87, **87**, 89, **91**, 198
Hotel, 108
Housing, 50-51, **100**, 102-108, 185, 190, 191, 193, **194**
  electricity, 98, 105, 107, 108
  plumbing, 107, 108
  rent, 102, 165
Howard, Tom, 11, **43**, 46, **49**, 134, 164-165, 182, 192
Hunting, 167, 170, 185-186

J:

Jackladder, 87, 89, **90**, **91**
Jobbers, 18, 25, 62-63, 113, 138
Jumpers, 71,

K:

Kiosk, 42, 161, 171, 175-178, 191, 194-201

L:

Labour relations, 134-135, 143,
Laferriere, Amedee, 52, **111**, 112, **125**, 143, **144**, 148, 178,
Logging by Rail in Algonquin Park, 8, 223
Logging camps, **24**, **25**, **59**, **64**, 65, 69-70, **69**, **70**, **79**, 149, 182
  food, 66-68, 69-70
Logging cars, 208
  brakes, 28, 77, 86, 208
Logging locomotives
  Climax, **34**
  Heisler, **33**
  saddle tank, 21, **21**, **38**
  railcars, 21, **21**, 208, 182
  Shay-geared, **20**, 21, **21**, **23**, **30**, **34**, **76**, **81**, **84**, 138, 157, **159**, **182**, **203**, 203-206, **204**, **205**, 208

Logging methods, 59-75, 99, 160
clearcutting, 13, 30
diameter limit, 60, 74-75
selection cutting, 200
sustained yield, 195
use of oxen, 31, **31**
Logging Railways, **32**
Fossmill Logging Railway, 50, 56, **64**, 77-87, **78-80**, 142-143
Nipissing and Nosbonsing Railway, 32, 36
Salmon River and Northern Railway, 16, 19, 22, **22-23**
South River Logging Railway, 30, **32**
Trout Creek Logging Railway, 31, **32**, 33, **33** **33**
Logging roads, 59, 64-65, 71-72
Logging sleighs, **70**, 72,
Logging tractors, 73-74, **74**, 195
Cletrac, 73, **198**
Logging trucks, **72**, 195, **196**
Lumber companies, other than the Fassett Lumber Corp., 148, 155, 196
Lumberyard, 94-95, **123**, 144-145, 160

M:
MacDonald, 51, **147**, 147-148, 163
Boyd, 54, 55, 118-119, 129, 146-148
Elva, 107, **125**, 185-187, 193
Gordon, 56, 136, 142-143
Myrtle, 113, 142-143, 147
Mackey, William, 172-173
McGibbon, 190
Jack, 11, 30, 42, **43**, 46, **48**, 59-60, **60**, 71, **109** 134, 163, 171
Stella 8, 21, 107,123, 135
Midwives, 114
Mill steam engine and boilers, **47**, 98, **98**, **164**
Montebello, 15, 101
Movies, 11, 126, 151
Music, 119, 126, 191

N:
North Bay Nugget, 137, 143, 144, 146-147, 151-152, 153-154, 157, 158, 182, 201
North Bay, 144, 148, 164, 166, 167, 193

O:
Orange Order, 56, 125, 143
Ottawa River, 15, 17, 28, 35
Our Chisholm Story, 143, 155, 179, 181

P:
Poaching, 183, 185-186
Post office,
Fassett, 18
Fossmill, 181, 184, 192
Powassan, 36, 108, 111, 114, 124, 126, 135, 147, 161, 186, 187, 189, 193
Prohibition, 127

R:
Radio, 107, 126, 183-184, 191
Railway station, **48**-49, **57**, 112-113, 152, 163, 175, 191
Railways, 31, **32**, 45,
wrecks, 152-155, 181-182, **181**
Recreation,
baseball, **10,** 11, **17**, 124
fishing, 75, 122-123, **122**
skating, **124**
swimming, 123-124, **123**
tennis, 124
Relief, 148-149, 165-166, **209**

S:
Sawmills, 9, 35-36, 93, 172
Fassett, Quebec, 9, 25-26
Fossmill, **44**, **47**, 56, **57**, 88-99, 163-164
Kiosk, **176**, 177, **197**
Schools, 116-121
discipline, 118, 120-121
Fassett, **18**, 19
Fossmill, **54**, 54-55, 115, **116**, 117-121, 168, 170
Kiosk, 191, **194**
music, 119-120
Separate, 179, 189
Wasing, 51-52, **52**, 168, 170, 189
Snow plow, **85**
Sprinkler tank, 72, **72**, **73**
Standard Chemical Company,
Fassett, Quebec, 15, 18, **27**, 30, 43
South River, Ontario, 30, 34, 151-152
Staniforth Lumber Company, 177, 195, 199
Staniforths, 8, 170, 199
Sydney, 17, **17**, **43**, 43, 47, 108-110, 161, 171, 175, 177, 188, 198-199, **200**

T:
Teachers
Bannister, Doris, 55-56, **55**
Chayer, Alma, 189, 191
Demeza, Joe, 120-122, **121**, **122**
Priest, Ted, 56, 115, 117-120, **118-119**, **120**
Smythe, Fred, 119-190, **120**
Telephones, 49, 140, 179, 184
Trapping, 185

U:
Unions, 71, 99, 151-152, 196, 198, **199**
International Woodworkers of America, 196
Woods Workers Union, 151
United States, 9, 18, 31, 46, 161, 171

W:
Wages, 69, 71, 98, 99, 111, 117, 121, 132, 135, 143, 151, 167, 177, 185, 196
Wasi Lake and River, 35, 36, 37, 41, 43, 46, 50, **78**, 122-123, 124, 129, 144, 155, 161, 183, 193
Wasing, 11, 50, 52, 125, 133, 166, 170, 179, 183, 189
White pine, 9, 15, 35, 42
Squared Timber, 59, 172, 175
Saw logs, 59, 172, 175
Women and girls, 116, 118, 122, 127, 187-188, 196, **197**
Workers Compensation, 69, 147, 152
World War Two, 34, 148, 188, 199

Logging and Railway Places to See, and Things to Do

Railway Destinations

The Timber Train – Mattawa, Ontario

The Timber Train runs between Mattawa, Ontario and Témiscaming, Quebec. It was named Ontario's new attraction of the year in 1999. The train travels 65km and back on a scenic route along the north shore of the Ottawa River, while guides provide information on the sights. An open air car and an improved convenience car were added in 1999. The train runs from May to October, Wednesday to Sunday. For a full schedule and reservations call (800) 648-4446 or visit: http://www.timbertrain.com

Iroquois Falls Home of the Shay Locomotive – Iroquois Falls, Ontario

Iroquois Falls has a wide variety of tourist attractions, but from the point of view of the railway buff their well preserved 70 ton three truck Shay locomotive, located next to the town's old train station, is a real treat. This particular Shay was used on the Standard Chemical Company Railway in the early 1940s (see page 34 of this book). For further information call (705) 232-4546.

The Komoka Railway Museum – near London, Ontario

This museum is located 8 miles west of London, Ontario and features a restored 1913 Shay locomotive as well as a wide variety of railway artifacts and a model railway. It is open from June through September and the $2 entrance fee is a real bargain. For further information call (519) 657-1912 or visit: http://www.komokarail.ca

The National Museum of Science and Technology – Ottawa, Ontario

As the name indicates, this Ottawa museum features many fascinating exhibits, including a Shay locomotive and railway equipment. Rides on the Shay are occasionally available under the direction of the Bytown Railway Society. For further information call (613) 991-3044 or visit: http://www.science-tech.nmstc.ca

Cass Scenic Railway State Park – West Virginia

The premier place to see and ride on Shay locomotives. Nestled in the mountains of West Virginia, the town of Cass has a country store, museum and train station, all of which can be explored before your trip on one of the Shay locomotives along their historic railway. For information write box 107 Cass, West Virginia 24927 or phone 1-800-CALL-WVA or visit: http://www.neumedia.net/~cassvr/

The South Simcoe Railway – Tottenham, Ontario

This memorable train ride from Tottenham to Beeton features restored 1920s coaches pulled by original CNR steam locomotives. The Tottenham Station has a gift shop and snack caboose. The train operates weekends from May to November, and more often in July and August. For information call (905) 936-5815 or visit: http:// www.steamtrain.com

Logging Destinations

Algonquin Park Logger's Museum – Algonquin Park

This open air walking tour features old logging equipment, such as an alligator and locomotive, old buildings and a log chute and dam. An information centre houses the park's second bookstore. The two mini-dioramas are very informative. An annual Logger's Day in July is a must for logging enthusiasts. For further information call (613) 637-2828 or visit: http://www.algonquinpark.on.ca

Atikokan Centennial Museum and Historical Park – Atikokan, Ontario

A Shay locomotive is on display as well as logging and mining equipment. For details call (807) 597-6585

The Canadian Ecology Centre – near Mattawa, Ontario

The CEC is located at the Samuel de Champlain Provincial Park, east of Mattawa, Ontario. The new centre combines forestry training, educational opportunities and eco-tourism in a natural and architecturally unique environment. For details call (888) 747-7577 or visit: http://www.canadian-ecology.com

Centennial Park 1910 Logging Museum – Thunder Bay, Ontario

The museum features a 1910 bush camp and a collection of artifacts from the 1900s. For information call (807) 625-2313 or fax (807) 625-3258.

The Hope Sawmill Museum – Peterborough, Ontario

This historic lumber mill was built in 1836 and restored at the Otonabee Regional Conservation Authority near Peterborough. An interesting collection of carpentry tools is on display at the Peterborough Centennial Museum. The Hope Sawmill museum is found 1km north of Lang, where the Lang Pioneer Village is also located.

Latchford Heritage Logging Attraction – Latchford, Ontario

This museum is on the site of the House of Memories Museum in Latchford. It features a Logger's Hall of Fame which includes a profile of Sydney Staniforth of the Fassett/Staniforth companies. It has an H-style bunkhouse and cookery, a sawmill, ice house and blacksmith shop. For further information visit: http://www.nt.net/latchford/log.htm

Martin River Provincial Park logging camp – Fifty-fire Km. north of North Bay on Highway 11.

The logging camp contains 8 log buildings and various pieces of logging equipment can be seen during camp hours. Tours are available Friday through Sunday in the summer. The camp can be seen without staying at the Park. For further information call (705) 892-2200 or Fax (705) 892-2142.

Muskoka Pioneer Village – Huntsville, Ontario
This 55 acre village of the 1860 era has a variety of pioneer exhibits including an old sawmill that saws logs for buildings in the village. Their steam train, the "Portage Flyer," will connect in the year 2000 with a turn of the century train station which will open on July 1, 1999. For further information call (705) 789-7576 or visit: http://muskokapioneervillage.org

Pennsylvania Lumber Museum – Potter County, Pennsylvania
This museum has restored the last remaining Barnhart log loader and has it on display along with a 1912 Shay locomotive, working sawmill, and logging camp. For information call (814) 435-2652 or visit: http://www.lumbermuseum.org

The Authors

Paul Mackey: Paul has studied history and anthropology and has extensive experience in heritage restoration. He is an illustrator/photographer and has experience as an historical interpreter. He has a special interest in social history and the history of science and technology, especially railways and ships.

Doug Mackey: Doug is a former teacher, principal and school superintendent and now lives in Chisholm Township near North Bay, Ontario, close to the former site of Fossmill. He has many years of experience in community development and as a writer and researcher. As a former museum curator and as a local historian, he has a special interest in the forest industry, logging and logging railways.

Fossmill today: *The authors take a group of scouts on a tour of Fossmill in 1998.*